The New Cosmos

The Omega Nebula in Sagittarius
Photo courtesy Canada-France-Hawaii Telescope Corp.

ASTRONOMY LIBRARY NO. 6

The Astronomy of Our Galaxy and Beyond

from ASTRONOMY Magazine

with contributions by Alan P. Boss, Kenneth Brecher, Robert Burnham, Jack O. Burns, Ken Croswell, David J. Eicher, Peter Jedicke, Jeff Kanipe, David H. Levy, Laurence A. Marschall, Richard Monda, Govert Schilling, Richard Talcott, Gerrit L. Verschuur, and Belinda J. Wilkes

Kalmbach Books
Waukesha, Wisconsin

Books in the Astronomy Library Series

The Moon; an observing guide for backyard telescopes, by Michael T. Kitt. (Astronomy Library No. 1) Kalmbach Publishing Co., Waukesha, Wisconsin, 1992.

Beyond the Solar System; 100 best deep-sky objects for amateur astronomers, by David J. Eicher. (Astronomy Library No. 2) AstroMedia, a division of Kalmbach Publishing Co., Waukesha, Wisconsin, 1992.

The Universe from Your Backyard; a guide to deep-sky objects from ASTRONOMY Magazine, by David J. Eicher. (Astronomy Library No. 3) AstroMedia, a division of Kalmbach Publishing Co., Waukesha, Wisconsin, 1992.

Stars and Galaxies; ASTRONOMY's Guide to exploring the cosmos, edited by David J. Eicher. (Astronomy Library No. 4) AstroMedia, a division of Kalmbach Publishing Co., Waukesha, Wisconsin, 1992.

***Galaxies and the Universe; an observing guide from* Deep Sky**, edited by David J. Eicher. (Astronomy Library No. 5) Kalmbach Publishing Co., Waukesha, Wisconsin, 1992.

The New Cosmos; the astronomy of our Galaxy and beyond. (Astronomy Library No. 6) Kalmbach Publishing Co., Waukesha, Wisconsin, 1992.

**FOR THE READERS OF ASTRONOMY MAGAZINE,
the best group of enthusiasts we know.**

 Published by Kalmbach Publishing Co., 21027 Crossroads Circle, P.O. Box 1612, Waukesha, WI 53187. Printed in Hong Kong.

Editorial staff: Richard Berry, David H. Bruning, Robert Burnham, Stephen Cole, Alan Dyer, David J. Eicher, Robert L. Hayden, Jr., Jeff Kanipe, Michael Emmerich, Kristine R. Majdacic, Christine Reel, Rhoda I. Sherwood, Tracy Staedter, Margaret Sullivan, and Richard Talcott.

Art director: Lawrence Luser
Cover design by Lisa Bergman

The material in this book first appeared as articles in ASTRONOMY magazine.
They are reprinted here in their entirety.

The New cosmos / from Astronomy magazine.
p. cm.
Includes bibiographical references and index.
ISBN 0-913135-15-1
1. Astronomy--Popular works. I. Astronomy.
QB44.2.N49 1992 92-20538

Contents

Our Galaxy

Galaxies and Beyond

Distorted Spiral Galaxy M94 in Canes Venatici
Photo courtesy Rudolf Schild, Harvard-Smithsonian Center for Astrophysics

Preface

As the 21st century approaches, astronomy is, once again, in ferment. New missions to the planets, sophisticated ground-based telescopes, space-based observatories, and improved observing techniques promise to repaint the picture of the universe that we now know. Much of what astronomers will learn over the coming years and decades will focus on answering "big picture" questions — did the Big Bang really happen? Will the universe expand forever? How old is our Galaxy and how did it form? What powers high-energy objects like quasars?

Astronomers have over the past 25 years sent a spacecraft to all of the planets except Pluto. The solar system is no longer the ultimate frontier. Now, using instruments like the Keck Observatory and the Hubble Space Telescope, astronomers' efforts will increasingly focus on fundamental research in our Galaxy, the Milky Way, and beyond.

The New Cosmos showcases material from ASTRONOMY magazine that represents the best and brightest areas of current research on our Galaxy and other galaxies. The editors of ASTRONOMY hope that this material, culled from the past three years of the magazine, will provide an exciting, accurate, up-to-date look at exactly what is going on today in professional astronomy. Open this volume and you'll see what astronomers are currently thinking about a multitude of subjects. Does intelligent life exist on the nearest star to our Sun, Alpha Centauri? How do double stars form? Will Supernova 1987A undergo an outburst several years from now? What role do clouds of wispy dust play in star formation? What is the most likely object in the Galaxy to harbor a black hole?

One of the most intriguing aspects of astronomy is that for every question solved, several more are raised. For example, in conducting straightforward mapping of the magnetic fields round the Milky Way, astronomers discovered magnetic "fossils" from jets of ionized gas that long ago poured out of the center of our Galaxy. These fossils are the remnants of ancient magnetic fields that show us how active our Galaxy was early in its life. From observations scattered across the entire electromagnetic spectrum, astronomers are piecing together the puzzle of quasars and hypothesizing that they are supermassive black holes surrounded by a thick ring of dust and gas. Drawing on numerous observations and considerable amounts of reasoning, cosmologists are proclaiming that cosmology itself — the study of the origin of the universe — is made possible by the very time in which we exist. And similar new vistas are opening up in every area of astronomical research.

In addition to providing a front row seat for astronomical research, *The New Cosmos* will take you on a ride through the very fabric of the universe. The first article, "Glorious Universe," which I wrote, proclaims that the evolving universe is a vast, magnificent canvas on which we find ourselves occupying just the smallest of corners — but it's a highly interesting corner all the same. We suspect you'll enjoy this sweeping tour of the astronomers' new cosmos, and hope that you feel a little closer to the grand universe that surrounds us.

Robert Burnham
Editor, ASTRONOMY magazine
Waukesha, Wisconsin
September 1992

GLORIOUS UNIVERSE

The evolving universe is a vast, magnificent canvas on which we find ourselves occupying just the smallest of corners — but it's a highly interesting corner all the same.

All the matter, space, and time there ever has been or ever will be popped into existence 10 to 20 billion years ago.

This tiny, supercompressed point — a "singularity" in the jargon of physicists — was unimaginably hot and packed with energy. It burst forth in a titanic explosion and, expanding and changing ever since, it has become the universe that we — and who knows who else? — call home. If some cosmologists are right, it's the only universe there ever has been, but there's no way of knowing for sure. All they can say is that the universe is expanding much too fast to stop. Like a solidly hit fast ball, the Big Bang is going — going — gone.

But there's much yet to happen before everything flies over the outfield fence. Whirling galaxies will spark star-fire from drifting dust motes and volatile gas. Generations of stars will live and die; solar systems will form and evolve; worlds will be born. The full-color spectacle of the universe at work will fill the night with light, even as the voids between the galaxies grow larger and darker.

Eventually, however, the grand machine will run down. Eons from now, the last stars will grow cold and the galaxies will disperse. But the story may still not be over — matter itself may continue to evolve, with seemingly stable atoms becoming senile and spontaneously disintegrating.

What is the ultimate fate of the universe? That question is as mysterious as where the Big Bang itself came from — and just as unanswerable.

BEGINNINGS

Every astronomer knows that as telescopes look deeper into the universe, they also look deeper into time past. Moonlight is 1.3 seconds old when it spills across the floor of your living room, while sunlight ages 8 minutes in the time it takes to travel to Earth. The soft light of the Great Galaxy in Andromeda, rising in the northeast these evenings, is over 2 million years old.

But the oldest light of all isn't even light as we think of it. It is a very faint radio hiss that comes from every direction, all at once. This "light" is the fading fireball of the Big Bang, the cooling fire of creation, and it poses

By Robert Burnham
Illustrated by Adolf Schaller

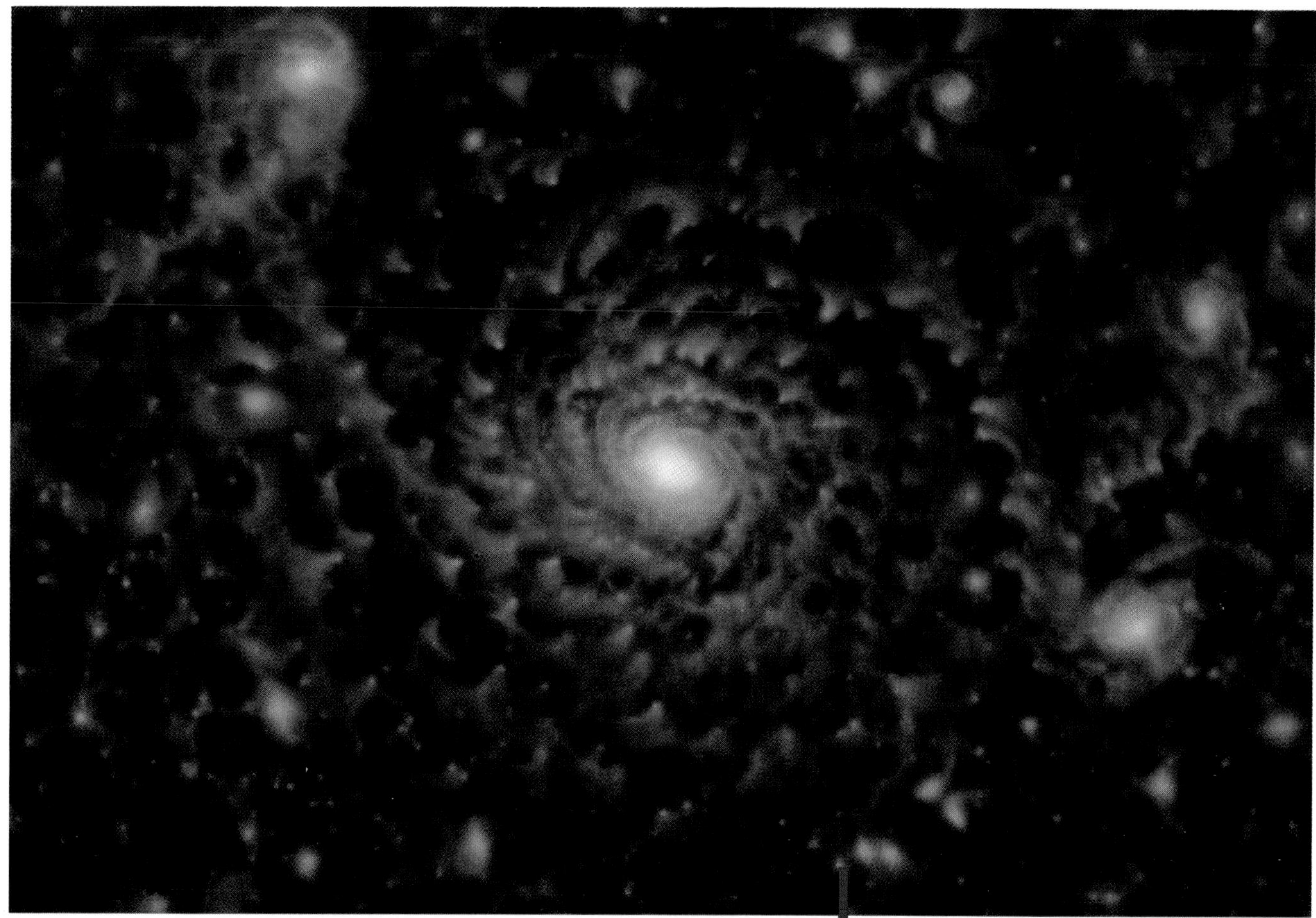

a problem for cosmologists because it is very, very smooth. No matter where they aim their instruments, the hiss sounds exactly the same. The smoothness of the hiss shows that the Big Bang was the tidiest, most well-mannered explosion that ever was — and yet soon after the Bang, galaxies appeared. Where did they come from? Gravity could build a galaxy quickly, but it has to have a seed to start with. The seed could be either a slightly denser clump of matter or a wrinkle in the explosion. Cosmologists argue over the observations, which seem to show that the early Big Bang had neither, and they construct elaborate theories from which galaxies spring forth instead out of such arcana as cold dark matter, false vacuums, or quantum tunneling.

But while cosmologists puzzle over how the universe made something out of (almost) nothing, the arrangement of galaxies also harbors unanswered questions. Why do they group into clusters, clusters of clusters, and even higher orders of organization? Every time astronomers push a survey to greater distances and encompass larger volumes of space, they find ever-larger structures — filaments, giant curving sheets and strings of galaxies, enormous yawning voids — until the very fabric of the universe seems packed with foamy soapsuds.

Other questions emerge when astronomers examine the most distant galaxies. Those nearer in time to the Big Bang look different from those close to Earth. The earliest galaxies seem to have strong central engines beating in their hearts: quasars, which fountain energy at rates the universe

HOW DOES A GALAXY GROW? *Astronomers do not yet understand how large organized structures like galaxies arose from an extremely smooth Big Bang.*

SHROUDED IN COSMOLOGICAL MYSTERY, *the time right after the Big Bang contained the seeds of everything — stars, nebulae, and galaxies — that has followed.*

THE COSMIC ARCHITECTURE *contains galaxies, clusters of galaxies, and clusters of clusters. Where do the congregations stop?*

has never seen since their heyday. Discovered thirty years ago, quasars are now thought to be black holes devouring material in the crowded centers of their host galaxies. Squeezed and heated by tidal forces beyond comprehension, the tortured matter flashes fierce radiation before it vanishes forever into the black hole.

Between the distant quasars and the stars of the Sun's neighborhood lies a vast middle ground where the architecture of the cosmos stands in full display. Far from scattering at random as astronomers once surmised, galaxies tend to congregate, as people do. Moreover, their congregations also congregate, and so on. These vast clouds of organization — in which galaxies behave like swarming dust motes — have taught astronomers that there's much more to the universe than meets the eye. An astronomer can examine a galaxy and from its brightness make a guess as to how many stars and how much dusty gas it contains. Yet when we track the movements of galaxies in a self-contained cluster, they move as if each galaxy were far heavier than it appears. It's like watching go-carts move as unstoppably as trucks. What is this mass that looms invisibly among the bright lights of the galaxies? No one knows.

The galaxies themselves pose puzzles. Why are there only two main kinds, spiral and elliptical? What determines which kind a galaxy will become? The Milky Way is a spiral galaxy; these account for about a third of the total. Spirals have central bulges with old reddish stars surrounded by spiral-armed disks filled with young stars and the dusty gas from which they were made. Ellipticals, on the other hand, are often larger but contain little except older yellow and red stars. Lacking the abundant gas to make

THE FIRST STARS AND THE GALAXIES *were made entirely of hydrogen and helium. Everything else, including the heavier elements we are made from, had to be created in this first generation of stars.*

THE ENDS OF STELLAR LIVES *are many. One such is the neutron star, the fate of most massive stars. If a neutron star is surrounded by a disk of gas and dust, it may flash into temporary brilliance as material falls onto its surface and detonates with multi-megaton force.*

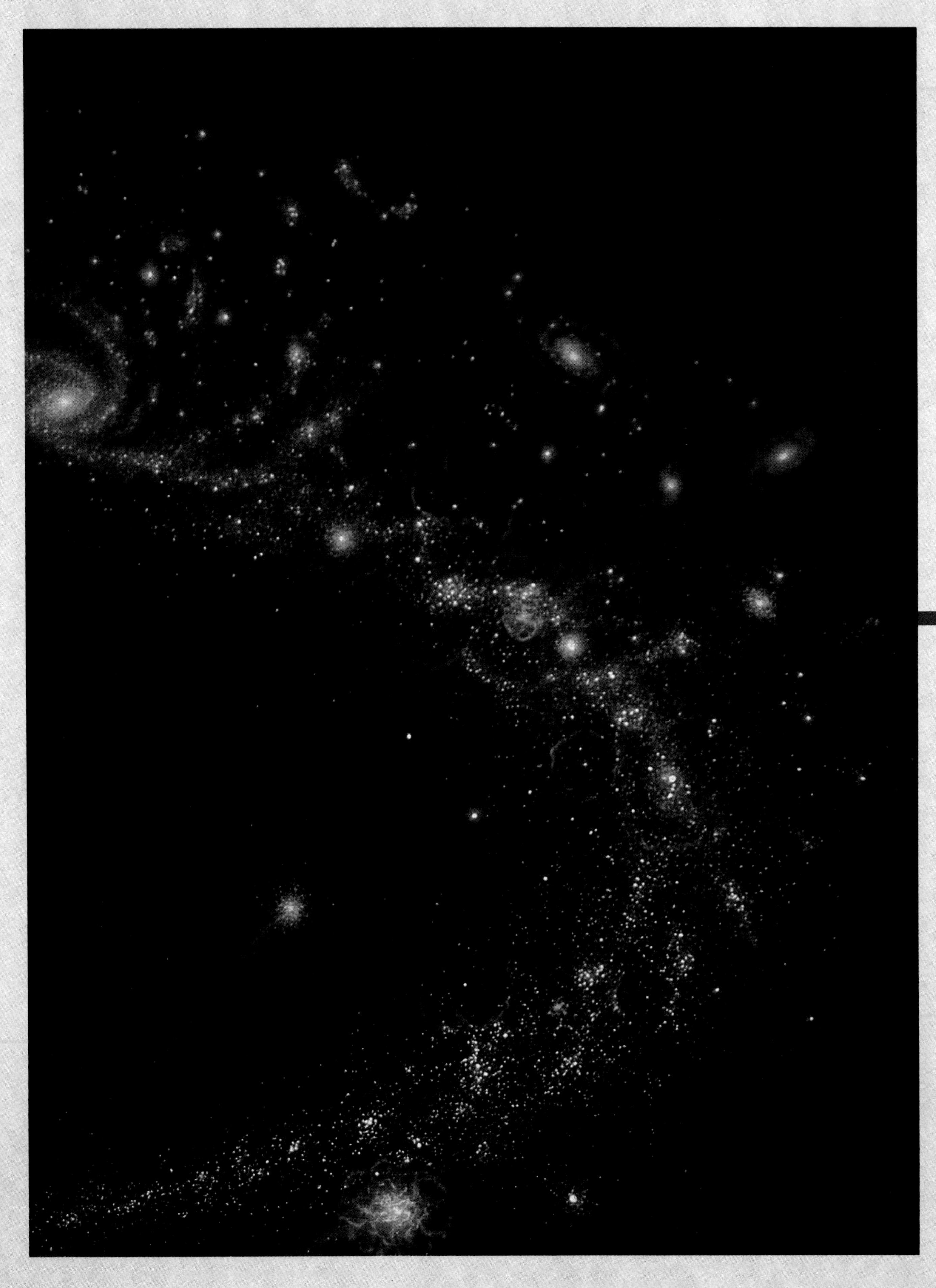

A FIERCE "WIND" STRIPPED *the inner solar system of gas and dust when the early Sun ignited its nuclear fires for the first time.*

A SHOCK WAVE MAY TRIGGER *stars to form in a spiral arm when it races through a cloud of galactic gas, dust, and molecules. This push is enough to start the cloud collapsing, and its largest lumps will form the first stars.*

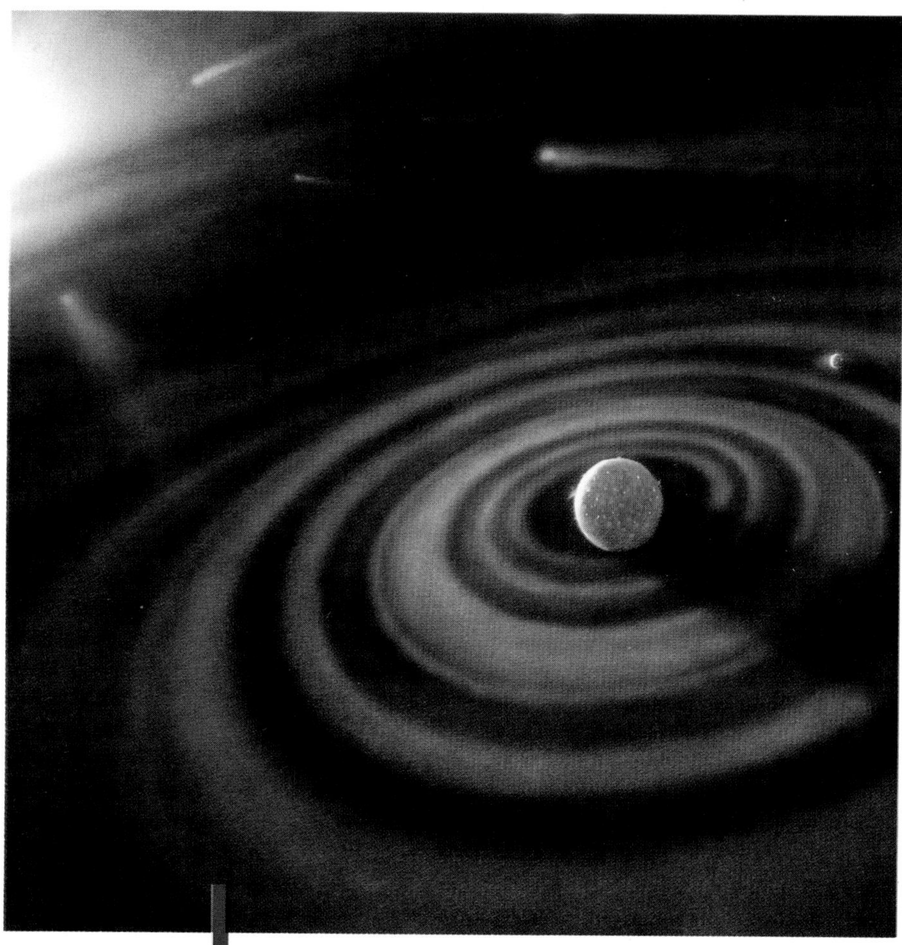

THE EARTH HAD A HOT AND VIOLENT BIRTH *at the heart of a cloud of debris which rained ceaselessly upon it for millions of years.*

more stars, these galaxies are doomed to a sterile future, growing older and colder with little hope of new star-life to rejuvenate them.

FACTORIES OF THE COSMOS

Spiral galaxies are nature's star-making machines. The seat of action in a spiral such as the Milky Way or the Andromeda Galaxy is the disk, the platter-shaped home of unnumbered stars and clouds of gas and dust. Sweeping round in a spiral pattern that endures for millions of years, shock waves in the disk periodically nudge the dust and gas. Jolted out of stasis, knots of gas begin to contract under gravity and grow warmer as their particles jostle one another. The knots eventually grow hot enough to light thermonuclear fires in their centers and become stars. Then, like new-lit lamps hung on a carousel, these drift onward through the night, orbiting the distant galactic center.

In time the larger stars race through their fuel and begin to die. Some become red giants — bloated gasbags whose outer envelopes are cool, filmy tissues of elements and molecules, easily lost to space. The biggest stars explode with titanic fury, throwing off their processed star-stuff into space in the wake of a

***THE EMERGENCE OF LIFE** was the most important event in Earth's history, but we know very little about the specific conditions under which it began.*

supernova. Ripping through dense clouds, the blast waves of supernovae may trigger new rounds of star-making. Our Sun is a second-generation star, born from stellar castoffs and hand-me-downs. It contains many elements — iron, oxygen, carbon, gold — that did not exist when the universe was new. To make the Sun, these elements had to be forged in the furnaces of older stars and broadcast back into space when those stars' lifelines drew to a close.

DEAD STARS' SOCIETY

Earth, Sun, and solar system were born together 4.5 to 5 billion years ago from an interstellar nebula of dusty gas. Laced with the debris of dead stars, the pre-solar nebula was parent to nearly everything in our lives, including warm sunlight, the blue sky overhead,

***MAKING PLANETS MAY BE ROUTINE** or it may be rare, but the recent discoveries of material around nearby stars encourages astronomers to think that solar systems may be fairly common in the Milky Way Galaxy.*

***A STAR IS NOT FOREVER** and our Sun is no exception. One day, evolution will turn it into a red giant. With a surface reaching past the orbit of Venus, the swollen Sun will devour its inner planets.*

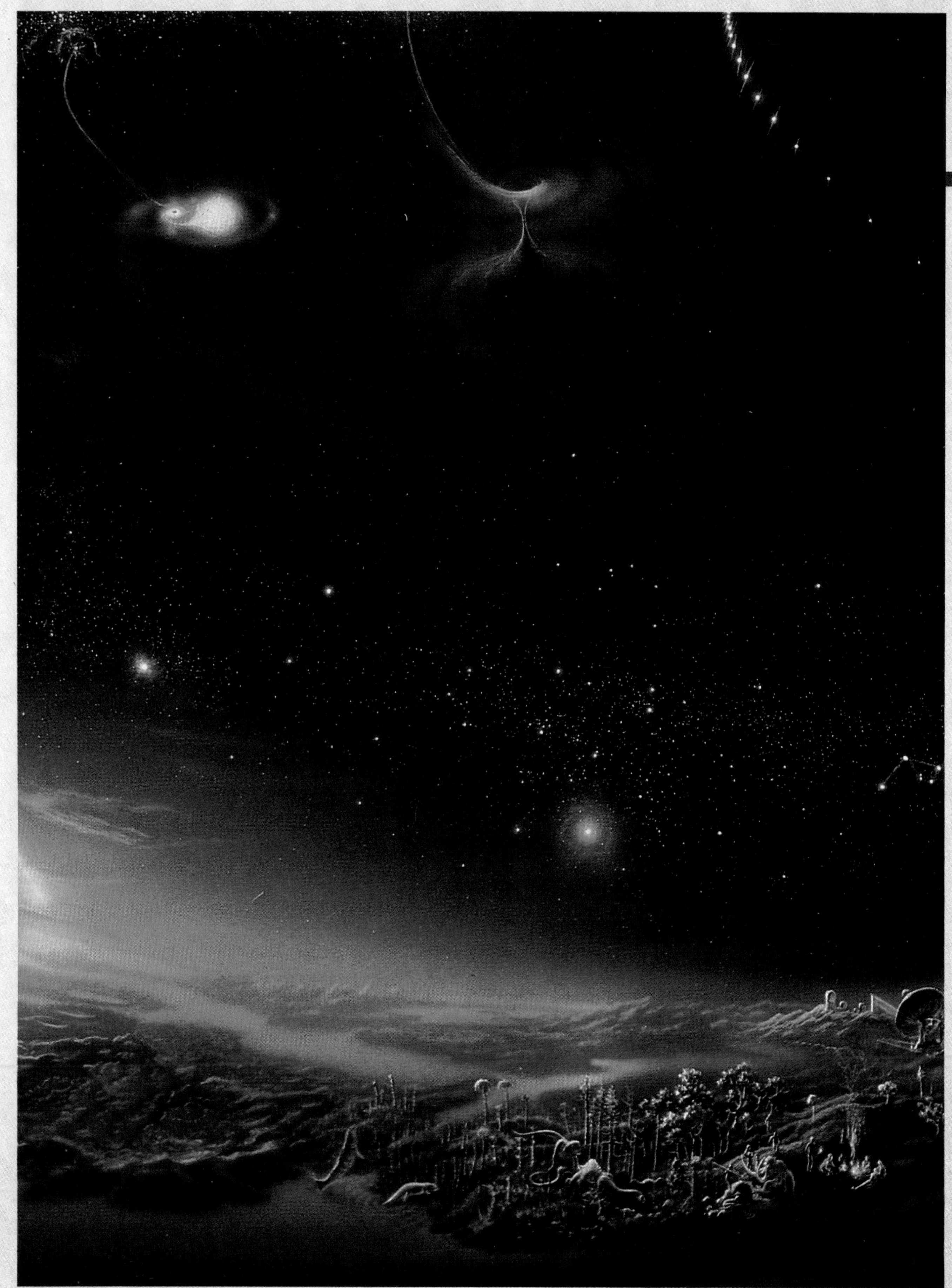

***EARTHLY LIFE THRIVES** wherever it finds a niche and this gives hope that we might find it elsewhere in the universe.*

and ourselves. Early Earth, however, was a hostile place — hot, repeatedly struck by asteroids, and shrouded in a choking atmosphere that would destroy us today. At a point perhaps 3.5 billion years ago, life began on Earth. Or it may have begun many times, destroyed by cataclysms each time but the last. Whether it formed in pools of water, as a skin on rocks, or in some other environment we can scarcely envision, life thrived and changed in response to its surroundings, always seeking a better metabolic fit between itself and the world. Where an organism's capabilities matched the demands of the environment, it reproduced and spread. Where the fit was poor, organisms died out.

Blindly, haphazardly, randomly chasing the paired grails of greatest reproductive success and least bodily insult, life evolved by fits and starts amid conditions that swung from the pleasantly benign to the immediately lethal. At least a dozen times, planet-wide ecological crises wiped out large fractions of all living organisms. Some crises may have been caused by the impacts of giant meteorites and others by reasons yet unknown. But in the aftermath of each crisis, life spawned wonderfully diverse assortments of new creatures. These quickly occupied newly vacant niches in the food chains and settled into a normal existence in which new species occurred relatively rarely.

FROM TRILOBITES TO TRITON

Many times in Earth's biological history, life has abruptly taken a step forward in complexity. The first step, after the origin of life itself, was the evolution of single-celled creatures. The second was the appearance of multicelled life, about a billion years ago. Other giant steps followed as millions of years rolled by. First trilobites, then the earliest land creatures, and fi-

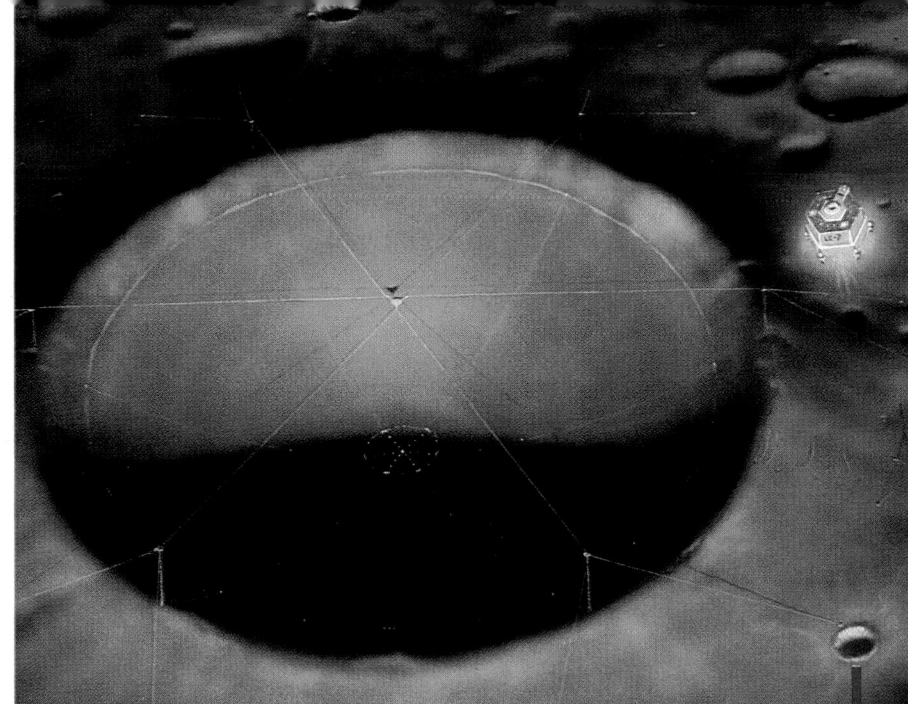

***A LUNAR ARECIBO RADIO TELESCOPE** on the Moon's farside could probe deeper into the cosmos, free from manmade interference.*

***DOES LIFE EXIST ELSEWHERE?** The solar system looks barren except for Earth, and yet scientists note that a drifting aerial form of life could exist in the warm cloud layers on Jupiter. And beyond the Sun? That's anyone's guess.*

nally giant reptiles appeared. The last major change, to mammalian-dominated life after the dinosaurs' demise, leaves us as the inheritors.

We pride ourselves on our intelligence and reason, which separate us from the other animals, though perhaps by less than we choose to think. Yet while it has made us, for now, the dominant species on the planet, there's no surety that our intelligence has survival value. Dinosaurs reigned supreme until a large asteroid struck 65 million years ago. In the chain of ecological catastrophes in its wake, life grew too difficult for dinosaurs and they perished. Yet how would we fare under similar conditions today? Could our technology cope?

Some space scientists have given serious thought to an idea called planetary engineering, or terraforming. This means taking one of the barren, hostile worlds of the solar system and altering its environment to permit life and perhaps human settlement. One example might be melting the Martian polar caps to release carbon dioxide and thicken the planet's atmosphere. This would provide greenhouse heating to warm the surface and raise the pressure to the point where liquid water could exist. Or, closer to home, we could give the Moon an atmosphere and hydrosphere, transforming it into a second, smaller Earth just next door. Another proposal would seed the clouds of Venus with organisms genetically engineered to reduce the massive carbon dioxide atmosphere and lower the surface temperature and pressure.

Some people are delighted by these scenarios, others horrified. Yet even if the plans never come to pass, it seems clear that should humans hold true to their heritage, we will continue to explore the worlds around us. Astronomy could be called the science of everything — and curiously, it's the only science that has so far escaped being linked to terrestrial woes. Perhaps this is true only because the one rule of the night sky seems to be "Look, but don't touch." For good or ill, this means that everyone sees the same universe and no one can stake a claim on it.

We walk outdoors on a cool, starlit night and look up into a beautiful universe about which we know so little. If we knew how to read it, nature's book could tell us all the secrets of the ages. Every day, astronomers and theorists decipher a couple more words or lines from the book, adding something to our knowledge. But there is so much to do and the nights are so few and so short. □

***WE HAVE CHANGED EARTH'S** environment largely by inadvertence, but what if we chose to alter another world? Could the Moon be made habitable with an atmosphere and hydrosphere? Will ocean waves break one day on the shores of lunar seas?*

***CURIOSITY AND A DRIVE TO UNDERSTAND** the cosmos have characterized humanity for as long as we can tell. Our telescopes and space probes are simply the most recent steps in a journey that began many thousands of years before Stonehenge.*

Does Alpha Centauri Have Intelligent Life?

Could the Sun's nearest stellar neighbor provide a home for life? The chances are surprisingly good.

by Ken Croswell

TWO SUNS STAND IN THE SKY over a hypothetical planet in the Alpha Centauri system, just over 4 light-years from Earth. The possibility of life there is highly conjectural, but judged by the standards of our own solar system, Alpha Centauri offers very good chances for planets, life, and perhaps even intelligence. Painting by Michael Carroll.

If astronomers understand how stars work, then the Alpha Centauri system offers two chances for terrestrial-type planet systems.

Alpha Centauri is a special star — not only because it is the closest stellar system to the Sun but also because it is one of the relatively few places in the Milky Way Galaxy that may harbor intelligent life.

Whereas most of the Galaxy's stars are dim red and white dwarfs that could not support life, Alpha Centauri is bright and warm like the Sun and provides abundant energy. Moreover, Alpha Centauri is sufficiently old that any life would have had time to evolve and produce intelligence.

Although we are its closest neighbor, we cannot yet say whether or not life exists around Alpha Centauri. In fact, we don't know if Alpha Centauri even has planets. Despite such uncertainties, we can examine Alpha Centauri to see whether it offers what intelligent life requires. The more we investigate Alpha Centauri, the more promising it looks.

The Layout of Alpha Centauri

The star we call Alpha Centauri is actually a triple system. Its brightest and warmest star is called Alpha Centauri A. It is a yellow star with a spectral type of G2, exactly the same as the Sun's. Therefore its temperature and color also match those of the Sun. Alpha Centauri A is slightly more massive and luminous, however: its mass is 1.09 solar masses and its brightness is 54 percent greater than the Sun's.

Alpha Centauri B, the second brightest star in the system, lies close to A. It is an orange star, cooler and smaller than the Sun. Its spectral type is K1 and its surface temperature is 5300 kelvins, some 500 K lower than the Sun's. The mass of Alpha Centauri B is 0.90 solar masses and the star's brightness is just 44 percent the solar value.

Visible only from latitudes south of about 25° north, the Alpha Centauri system lies 4.35 light-years from the Sun. The two brightest stars orbit each other every 80 years. The mean separation between them is 23 astronomical units — similar to the distance between the Sun and Uranus. (One AU is the distance between the Sun and Earth.) However, the stars' orbit is eccentric, so the separation varies during their 80-year period. When closest, Alpha Centauri A and B lie 11 AU apart — the distance between the Sun and Saturn. When farthest, though, the two stars are 35 AU apart — the separation between the Sun and Neptune.

The third and faintest member of the Alpha Centauri system, Alpha Centauri C, lies a long way from its two brighter companions, in distance and in nature. In fact, Alpha Centauri C lies 13,000 AU from A and B, or 400 times the distance between the Sun and Neptune. This is so great a distance that some astronomers believe the star is not even gravitationally bound to A and B. If Alpha Centauri C *is* attached, however, it must take about a million years to orbit them. So great is the separation that Alpha Centauri C actually lies measurably closer to us than the other two: A and B are 4.35 light-years distant, but Alpha Centauri C is only 4.22 light-years away. Technically speaking, therefore, Alpha Centauri C is the nearest individual star to the Sun. Because of this proximity, astronomers call the star Proxima Centauri.

Proxima is quite far from A and B in other ways as well. Whereas Alpha Centauri A and B are stars like the Sun, Proxima is a dim red dwarf — much fainter, cooler, and smaller than the Sun. Proxima Centauri's spectral type is M5, its temperature is half the Sun's, its mass is one-tenth the Sun's, and its brightness is a mere 0.006 percent that of the Sun. Proxima is so faint that astronomers did not even discover it until 1915.

The Rarity of Alpha Centauri

Alpha Centauri is special because our Sun is special. The Sun emits lots of energy, just what life needs. But contrary to what many astronomy textbooks say, the Sun is *not* an average star. It is far above average in luminosity, mass, and size. If it weren't, we wouldn't be here.

We cannot appreciate either Alpha Centauri or the Sun until we look at the Galaxy's other stars. Most are incapable of sustaining life. Seventy percent of all stars in the Galaxy are red dwarfs like Proxima Centauri, too faint, too cool, and in some cases too variable to support life. About 15 percent are orange K-type dwarfs. Although the more luminous K dwarfs (like Alpha Centauri B) may be bright and warm enough for life, the fainter ones in the class may be too dim and cool. Another 10 percent of stars are white dwarfs — dying stars that either could not have life or must have destroyed any life they once had.

That leaves the brightest 5 percent of all stars in the Galaxy, a privileged group to which both the Sun and Alpha Centauri A belong. Most of this upper 5 percent consists of yellow G-type stars that are bright, warm, and good for life.

Because they power any life around them, the stars in a stellar system are the first things to consider for intelligent life. A star must pass five different tests before we can call it a promising place for life. Passing all five tests is not easy. Most stars in the Galaxy

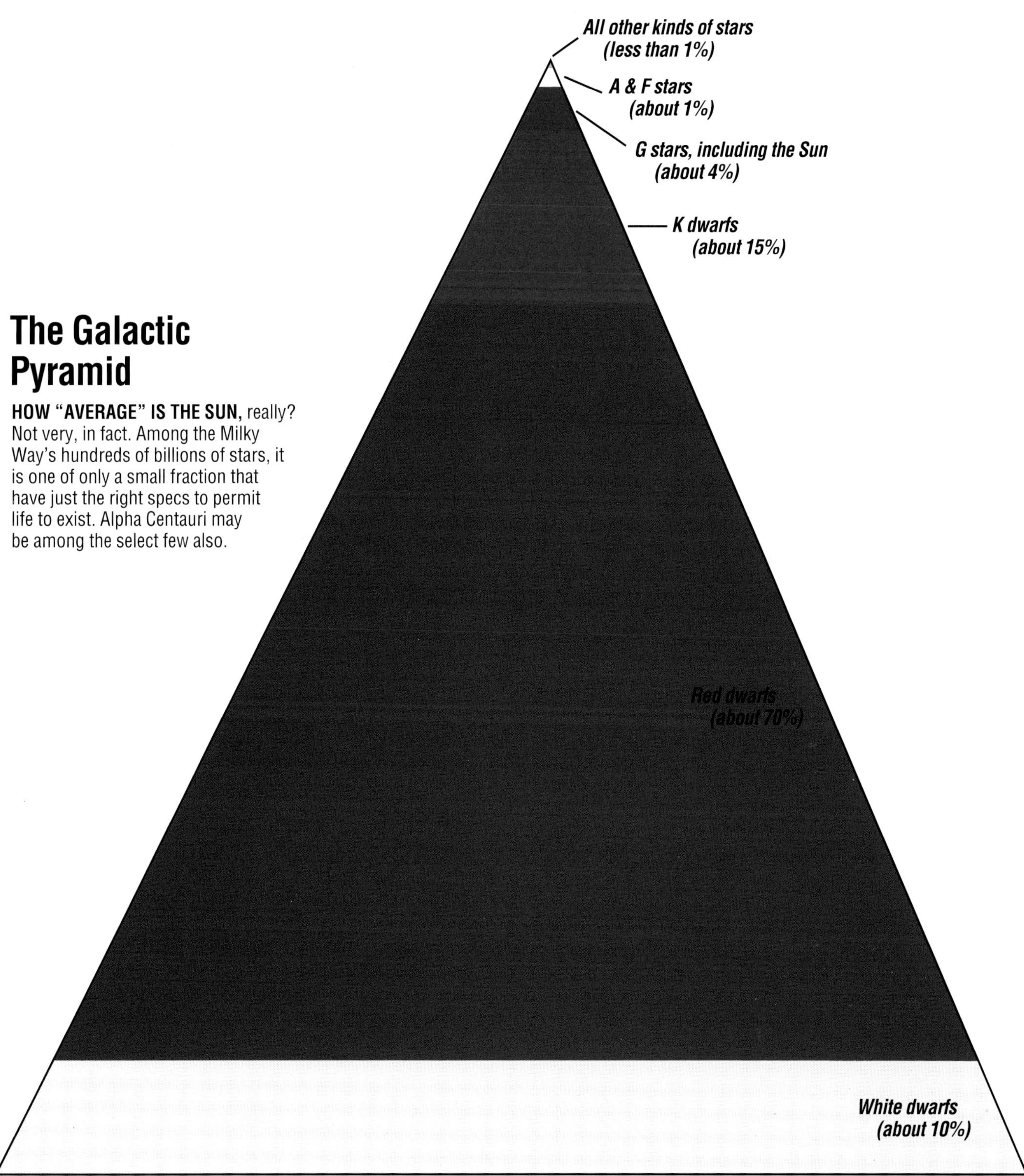

The Galactic Pyramid

HOW "AVERAGE" IS THE SUN, really? Not very, in fact. Among the Milky Way's hundreds of billions of stars, it is one of only a small fraction that have just the right specs to permit life to exist. Alpha Centauri may be among the select few also.

would fail. In the case of Alpha Centauri, however, we will see that Alpha Centauri A passes all five tests, Alpha Centauri B passes either all or all but one, and only Proxima Centauri flunks out.

Good Stars

The first test is the easiest. To ensure a star's maturity and stability, it has to be on the main sequence. Main-sequence stars fuse hydrogen into helium at their cores, generating light and heat. Because hydrogen is so abundant in stars, most of them stay on the main sequence a long time, giving life a chance to develop.

After a star exhausts the hydrogen at its core, the star goes through rapid changes, first becoming a red giant and then a white dwarf. These changes are fun for astronomers to study but highly detrimental to any life orbiting the star! Ninety percent of the Galaxy's stars are on the main sequence, so most, including

The Sun and Its Nearest Neighbors

	Sun	Alpha Cen A	Alpha Cen B	Proxima
Color	Yellow	Yellow	Orange	Red
Spectral type	G2	G2	K1	M5
Temperature	5800 K	5800 K	5300 K	2700 K
Mass	1.00	1.09	0.90	0.1
Radius	1.00	1.2	0.8	0.2
Brightness	1.00	1.54	0.44	0.00006
Apparent magnitude	–26.7	0.0	1.4	11.1
Absolute magnitude	4.85	4.38	5.74	15.54
Distance (light-years)	0.00	4.35	4.35	4.22

the Sun and all three components of Alpha Centauri, pass this test.

The second test is much tougher, however, and most stars in the Galaxy fail it. We want the star to have the right spectral type, because this determines how much energy a star emits. On the main sequence, the "earlier" the spectral type — that is, the hotter and bluer the star — the more energy it puts out.

Although they might at first seem the best, the hottest stars — those with spectral types O, B, A, and early F — are actually the worst. These stars are so bright they burn their fuel fast and die quickly. A typical O star like Iota Orionis has a lifetime of just 10 million years. A typical B star like Regulus lives just 100 million years. And a typical A star like Sirius lives only a billion years or so. On Earth, intelligence took 4.6 billion years to develop. If we are typical, then O, B, A, and early F stars can't have intelligent life because they burn out before it establishes itself.

The opposite end of the main sequence is just as bad, but in a different way. M-type main-sequence stars (red dwarfs such as Proxima Centauri) indeed live for tens or hundreds of billions of years. But they emit little energy during their long lives. Red dwarfs therefore do not give off enough energy to support life.

Between the stars that are too hot and those that are too cool, we find the stars that are just right. As our existence proves, yellow G-type stars like the Sun can give rise not only to life but to intelligent life. What about other spectral types? Late (cool) F stars may be fine, too — they produce even more energy than the Sun, and they probably live long enough to foster intelligence. More problematic are the orange K stars, cooler than the Sun. Do they produce enough energy? The brighter and hotter ones — K0, K1, and K2 stars — may, but the fainter and cooler K stars (later than K5) may not.

How does Alpha Centauri fare in this second test? With a G2 spectral type identical to the Sun's, Alpha Centauri A passes easily. Alpha Centauri B is a K1 star, so it is brighter and hotter than most K stars and therefore may pass this test or it may not. And, of course, the red dwarf Proxima Centauri is a hopeless case.

For our third test, a star must demonstrate that it is stable. We do not want the star's brightness to vary so much that the star would alternately freeze and fry any life that does manage to develop around it. Like the Sun, Alpha Centauri A and B are stable and do not vary in brightness.

But because Alpha Centauri A and B form a binary pair there's a further issue. How much does the light received by the planets of one star vary as the other star revolves around it? During their 80-year orbit, the separation between A and B changes from 11 AU to 35 AU. As viewed from the planets of Alpha Centauri A, the brightness of Alpha Centauri B increases as B approaches A and then decreases as B recedes from A. Likewise, as viewed from the planets of Alpha Centauri B, the brightness of Alpha Centauri A increases and decreases as the star approaches and recedes.

Fortunately, the variation is too small to matter. As viewed from a planet orbiting A, Alpha Centauri B ranges from apparent magnitude –18.1 at faintest to –20.6 at brightest. (The Full Moon shines at –12.7 magnitude and the Sun –26.7.) But even when B is brightest, Alpha Centauri A (viewed from one AU) is over 400 times brighter. B's variation therefore means nothing since the light received on Alpha Centauri A's planet comes almost entirely from its own star.

Reversing the scenario brings us to the same conclusion. When nearest B, Alpha Centauri A shines at magnitude –21.9 as viewed from the planets of Alpha Centauri B. When farthest, Alpha Centauri A drops to magnitude –19.4. But again, the light that B's planets receive is dominated by light from B, not A, so the variation isn't harmful. In this case, the total light received by a planet one AU from Alpha Centauri B varies by just 3 percent as A approaches and recedes. This is smaller than the variation we suffer each year as the Earth revolves about the Sun on its slightly eccentric orbit. So both A and B pass the test for stability.

But poor Proxima. It flunked our last test and it fails this, too. Like many red dwarfs, Proxima Centauri is a flare star, prone to outbursts that cause its light to double or triple in just a few minutes. Any life around Proxima would have to be quite hardy to survive such outbursts.

The fourth test concerns the stars' ages. The Sun is about 4.6 billion years old, so on Earth intelligent life took that long to evolve. This is a lengthy time and

The Local Top Ten Stars (of the Nearest 100)

Rank	Star	Absolute Magnitude	Spectral type	Distance (l-y)
1	Sirius A	1.42	A1	8.7
2	Altair	2.24	A7	16.5
3	Procyon A	2.64	F5	11.4
4	**Alpha Centauri A**	**4.38**	**G2**	**4.35**
5	Eta Cassiopeiae A	4.61	G0	19.1
6	Delta Pavonis	4.76	G7	18.6
7	**Sun**	**4.85**	**G2**	**0.0**
8	Tau Ceti	5.72	G8	11.8
9	**Alpha Centauri B**	**5.74**	**K1**	**4.35**
10	70 Ophiuchi A	5.76	K0	16.1

roughly half the stars in the Galaxy are younger than the Sun. If intelligence takes 4 to 5 billion years to develop, these young stars cannot have intelligent life around them yet. In assessing a star's chances for intelligent life, we therefore want the star to be at least as old as the Sun. For most stars, however, determining age is tough work. We know the Sun's age only because we can date rocks on the Earth and Moon.

Remarkably, though, we can also date Alpha Centauri. A main-sequence star slowly brightens as it ages. For example, the Sun is now about 40 percent brighter than it was at birth. If we know a star's mass precisely, we can calculate how the star's luminosity varies as the star ages. Alpha Centauri A and B form a binary, so we can determine their masses quite accurately. Because they are so close to us, we also know their distances and luminosities precisely. We can therefore determine how old these stars — with masses of 1.09 and 0.90 Suns — must be to be as bright as they now are. Brian Flannery of Exxon Corporation and Thomas Ayres of the University of Colorado did this in 1978 and found that Alpha Centauri A and B are both about 6 billion years old. In 1986, Pierre Demarque, D. B. Guenther, and William van Altena of Yale University used more recent data to find that both stars are about 5 billion years old. Alpha Centauri A and B are therefore a bit older than the Sun, so intelligence has had time to evolve around them.

What about Proxima? If Proxima was born with A and B, then it too must be 5 or 6 billion years old. However, the star's flare activity suggests instead it is young, since flare activity is a sign of youth. Proxima Centauri may be only a billion years old, which means it did not form with A and B but was captured by them. If Proxima really is only a billion or so years old, then it fails this test too.

Our fifth and final test: Do the stars have the heavy elements — such as carbon, nitrogen, oxygen, and iron — that life needs? Like most stars, the Sun is primarily hydrogen and helium, but 2 percent of the Sun's weight is in metals. (Astronomers call all elements heavier than helium "metals.") Although 2 percent may not sound like a lot, this enrichment of the material the Sun and planets formed from was enough to give rise to us. Furthermore, the Sun is metal-rich compared with many other stars in the Galaxy. In particular, older stars tend to have fewer metals. Since Alpha Centauri A and B are older than the Sun, we might worry that these stars do not have enough metals to develop life.

Fortunately, Alpha Centauri A and B are both metal-rich stars. Astronomers have measured the metal abundance of the two main stars in Alpha Centauri and they find that it has even more metals than the Sun does. Thus Alpha Centauri passes this test, too.

How did the three stars do in these tests? Alpha Centauri A passed all of them quite well. It is a stable, metal-rich, main-sequence G star with a slightly greater age than the Sun. Alpha Centauri B passed at least four of the five tests. It, too, is a stable, metal-rich, main-sequence star that is old enough for intelligent life. The only possible problem is its K spectral type, which indicates the star may be too faint and cool to support life. Finally, the red dwarf Proxima Centauri didn't do well at all.

Good Planets?

Vital as it is for life, a good star is not enough. We also need a good planet to orbit the good star — a warm, rocky planet like Earth, full of liquid water.

Unfortunately, it is here that big uncertainties arise. We don't know whether Alpha Centauri even *has* planets, much less whether they would be good places for life. Another problem is that because Alpha Centauri A and B lie close to each other, one star's gravity may disturb the planets orbiting the other star.

If Alpha Centauri A and B were very close together, then planets could revolve at a distance around both stars. Inhabitants of such a solar system would always see a double sun in their daytime sky. On the other hand, if the two stars were very far apart, then each star could have its own planetary system. Inhabitants would call one star their sun and see the other as a very bright point of light in the sky. Unfortunately, Alpha Centauri A and B fall between these two extremes, which poses problems for planets. (Since it is so far from its partners, Proxima Centauri can and

Possible planets at Alpha Centauri

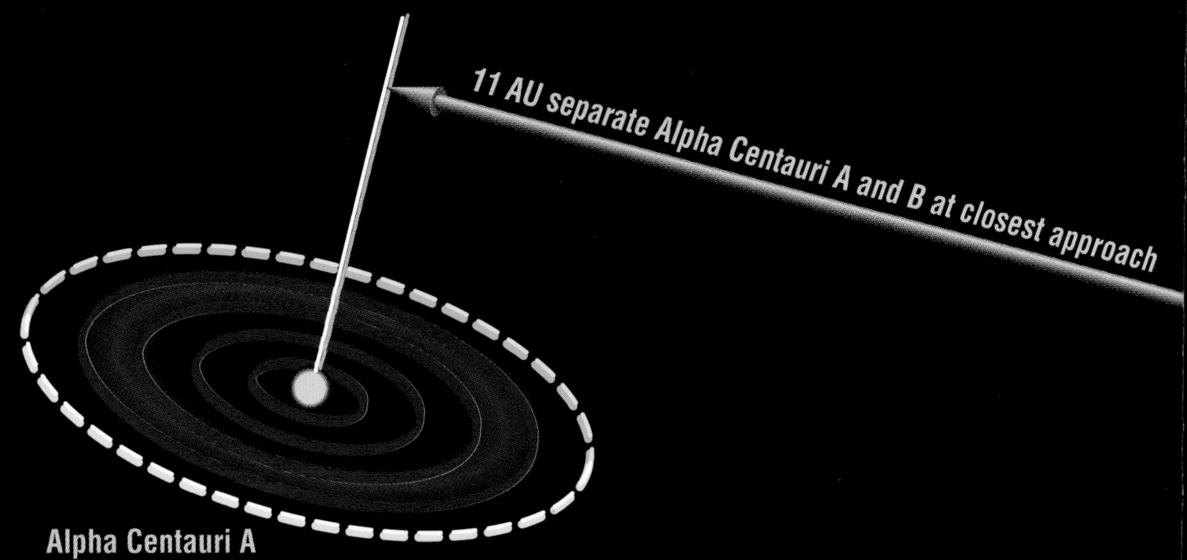

IF ALPHA CENTAURI A AND B HAVE PLANETS spaced apart like Mercury, Venus, Earth, and Mars, the chances are good that at least one of them will fall into the star's life zone (green band) the bright, warm region where water could remain liquid.

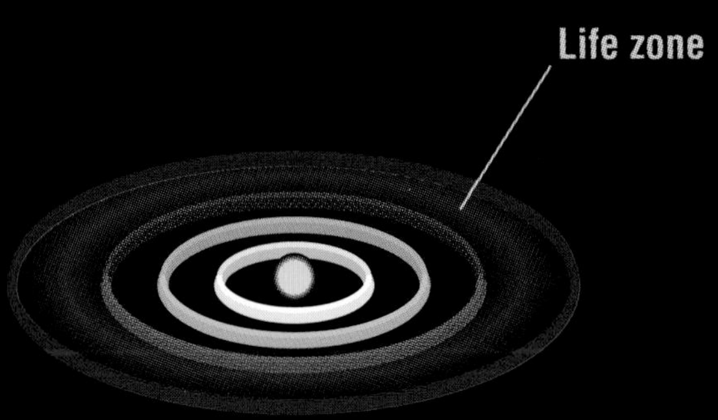

The Sun and its terrestrial planets (on the same scale)

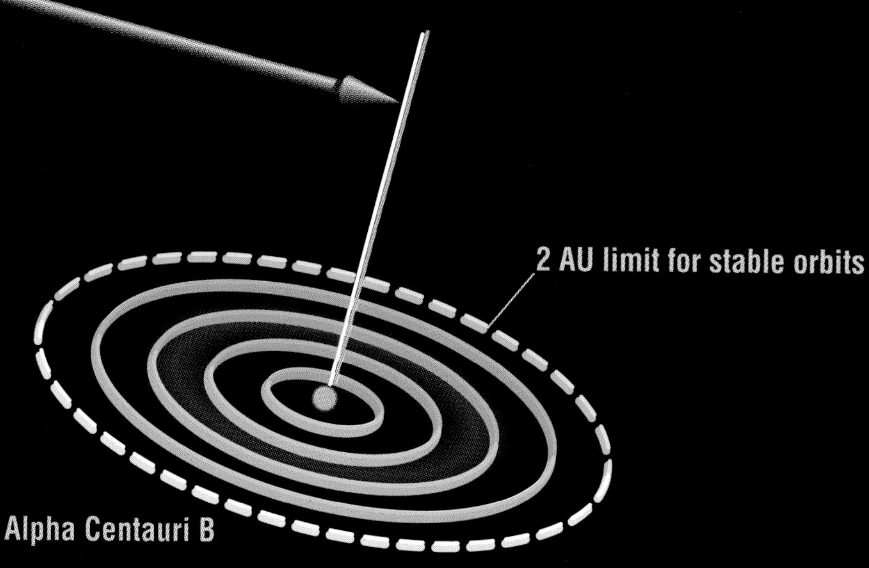
2 AU limit for stable orbits
Alpha Centauri B

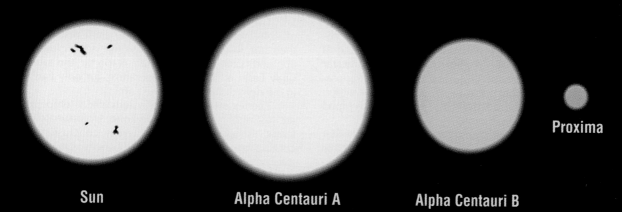

In true relative scale
Sun
Alpha Centauri A
Alpha Centauri B
Proxima

Ten Questions for Any Star

	Sun	Alpha Cen A	Alpha Cen B	Proxima A
On the main sequence?	Yes	Yes	Yes	Yes
Of the right spectral type?	Yes	Yes	Maybe	No
Constant in brightness?	Yes	Yes	Yes	No
Old enough?	Yes	Yes	Yes	No?
Rich in metals?	Yes	Yes	Yes	?
Has stable planetary orbits?	Yes	Yes	Yes	Yes
Could planets form?	Yes	?	?	Yes
Do planets actually exist?	Yes	?	?	?
Small, rocky planets possible?	Yes	Yes	Yes	Yes?
Planets in the life zone?	Yes	Maybe	Maybe	No

probably does have planets, but since Proxima is a red dwarf, the planets must be lifeless.)

Nonetheless, we can proceed as before by submitting the system to different tests. The first test: Do stable orbits exist? If we put a planet in orbit 1 AU from Alpha Centauri A, would it stay there? Or would the gravity of Alpha Centauri B rip it away? Celestial mechanics shows that stable orbits definitely exist around either star — provided the planets lie no farther from their star than one-fifth the closest distance the two stars ever get to each other. Since A and B get as close as 11 AU, this means that each star could have planets with stable orbits out to a distance of at least 2 AU.

In our solar system, four planets lie within 2 AU of the Sun: Mercury (0.4 AU from the Sun), Venus (0.7 AU), Earth (1.0 AU), and Mars (1.5 AU). Therefore, Alpha Centauri A and B could each have four planets with orbits just like those of the inner solar system.

The second test is tougher. Even though planets can exist around Alpha Centauri A and B, could such planets have formed in the first place? In the solar system, Jupiter's mighty gravity disturbed the developing solar system and prevented a planet from forming between it and Mars. That is why our solar system has an asteroid belt between Mars and Jupiter. The situation around Alpha Centauri could be far worse, for both Alpha Centauri A and B are about a thousand times more massive than Jupiter. When the protoplanetary material around Alpha Centauri A and B was attempting to form planets, the gravity of one star may have so thoroughly disturbed the developing planets of the other that no planets formed around either. As a consequence, all that may exist around Alpha Centauri A and B are multitudes of asteroids.

The ultimate question, though, is not whether planets have stable orbits or whether they could have formed, but whether they exist *today.* We have no evidence for any, of course, but some astronomers hope that when the *Hubble Space Telescope* is repaired it may spot planets around Alpha Centauri. Until then, we simply do not know if Alpha Centauri has any planets. We can nevertheless proceed by supposing each star does have four planets with orbits similar to those of Mercury, Venus, Earth, and Mars. The next question: Are these planets good or bad? "Good planets" are small, rocky worlds where life could form if the planets are the right distance from their star. "Bad planets" are huge gas-balls with lots of hydrogen and helium, probably incapable of forming advanced life even if they orbit close to their star. In our solar system, the good planets — Mercury, Venus, Earth, and Mars — lie in the inner solar system, whereas the bad planets — Jupiter, Saturn, Uranus, and Neptune — lie in the outer solar system.

Astronomers believe other solar systems may follow the same pattern: the inner planets small and rocky, the outer planets large and gaseous. Our planets condensed at different places and therefore at different temperatures in the primordial solar nebula. In the inner solar system (which was hot), only hardy substances like rock and iron could condense, which is why the four inner planets consist of rock and iron. In the outer solar system, however, the pervasive cold allowed substances other than rock and iron to survive, such as methane and ammonia. These planets grew huge and became the gas giants.

What do we expect around Alpha Centauri A and B? Any planets within two AU of either star must have formed from a hot nebula. These worlds, if they exist, are almost certainly small, rocky planets like Earth rather than lifeless gas giants like Jupiter. Thus even though both stars can have only truncated solar systems, any planets that did arise are just the sort we're looking for.

The final question we ask is whether any of the planets around A or B lie at the right distance from their star. In our solar system, Venus is too close to the Sun and is too hot, while Mars is too far and too cold. Only Earth is the right temperature, for only Earth lies at the right distance from the Sun.

The region in our solar system where temperatures are just right is called the life zone. Unfortunately, different astronomers reach different conclusions concerning how wide the Sun's life zone is. One estimate made many years ago says it extends from only 0.95 to 1.01 AU. A more recent estimate, however, says it extends from 0.95 to 1.5 AU. Of course, it is more than just a planet's distance that determines temperature. In general, the bigger a planet, the warmer it will be. In

Life at Alpha Centauri? The hardest questions to answer aren't astronomical — the real stumpers are the biological ones.

our solar system, Mars is small and cold, but some astronomers believe that if Mars were the size of Earth, it might be warm enough for life.

What is the chance that Alpha Centauri A or B has a planet in its life zone? In our system, inner planets are spaced about 0.4 AU apart. Suppose the planets of Alpha Centauri A and B are also spaced 0.4 AU apart. If the life zone of each star is 0.1 AU wide, then each star has a one in four chance of having a planet in its life zone. If the life zone is wider, the probability grows. For example, if the life zone is really 0.2 AU wide, then each star has a 50-50 chance of putting a planet into its life zone. Such a planet could be a warm, watery world like Earth and an excellent place for breeding life.

It's frustrating, but we can't reach any firm conclusions about Alpha Centauri. Whereas we could say with certainty that at least one of its stars is ideal, we can say little about any planets there. On the positive side, Earth-like planets can have stable, Earth-like orbits around either A or B. On the negative, such planets might never have formed in the first place and we have no evidence that planets of any kind exist around Alpha Centauri.

Does Alpha Centauri Have Intelligent Life?

But if Earth-like planets exist, did life actually develop on them? And is some of that life intelligent? Unfortunately, the answers depend not on astronomy but on biology. In confronting astronomical issues, we could often answer the questions we posed. The same is not true for the biological issues. In fact, we have no good answers *whatsoever* when we begin to ask biological questions. Yet the question of intelligent life around Alpha Centauri may hinge more on biological factors than on astronomical ones.

There are three chief biological questions. The first: Given a warm, watery planet like Earth governed by a bright, warm star like the Sun, will primitive life develop? Scientists sometimes assume it will since that is what happened on Earth. But we have no assurance that Earth is typical, for it is the only planet on which we have ever detected life. Furthermore, the origin of life even here on Earth remains a mystery. Thus even if Alpha Centauri has warm, watery planets, these worlds could be completely lifeless.

Second, if primitive life does manage to develop, will that life evolve into more advanced forms? On Earth it obviously did. Single-celled organisms became multi-celled, which then evolved into more complex life forms such as plants and animals. But we have no guarantee that the same process will occur on other worlds. Even if Alpha Centauri has warm, watery planets bearing life, it's possible the life there consists of nothing more than primitive single-celled organisms.

Finally, if advanced life does develop, will intelligent life evolve with it? Again, no guarantees. On other planets, strength and speed may count for more than intelligence. Even if a species like us managed to arise on such a planet, we might soon be killed off by stronger, faster, dumber competitors. Thus Alpha Centauri could have planets with advanced life that lacks intelligence. The dominant species might be a plant or a tree.

The Promise of Alpha Centauri

The three biological questions are not unique to Alpha Centauri. They confront us whenever we examine any star system for intelligent life. That we cannot answer them should not detract from the great promise that Alpha Centauri holds for harboring intelligent life.

Alpha Centauri A is one of the best stars we could ask for — it resembles the Sun more than any of the 70 nearest stars do. With a spectral type of G2, a high metal abundance, and an age exceeding the Sun's, the star could easily support intelligent life on its planets, if such exist. And Alpha Centauri B, though fainter and cooler than the Sun, also offers hope that life has established itself on planets orbiting it.

Furthermore, these very promising stars are not just any stars in the Galaxy. They are the stars closest to the Sun. The next nearest star to us beyond Alpha Centauri is Barnard's star, a worthless red dwarf like Proxima. The next closest G-type star beyond Alpha Centauri is Tau Ceti, twice as distant as Barnard's star. Though it is a single star that may have planets, Tau Ceti has only about a third the Sun's abundance of the heavy elements such as carbon and oxygen that life needs.

Alpha Centauri is therefore very special — special not only for its proximity but also for its promise. If Alpha Centauri has planets, it is even more special since one or more of those planets could resemble Earth. And if we ever launch missions to other stars in the Galaxy, Alpha Centauri will certainly be our first target. □

Ken Croswell received his doctorate in astronomy from Harvard University and has written for New Scientist, Star Date, *and* Time-Life Books. *His last article for ASTRONOMY was "A Star That Breaks All the Rules" in the January 1991 issue.*

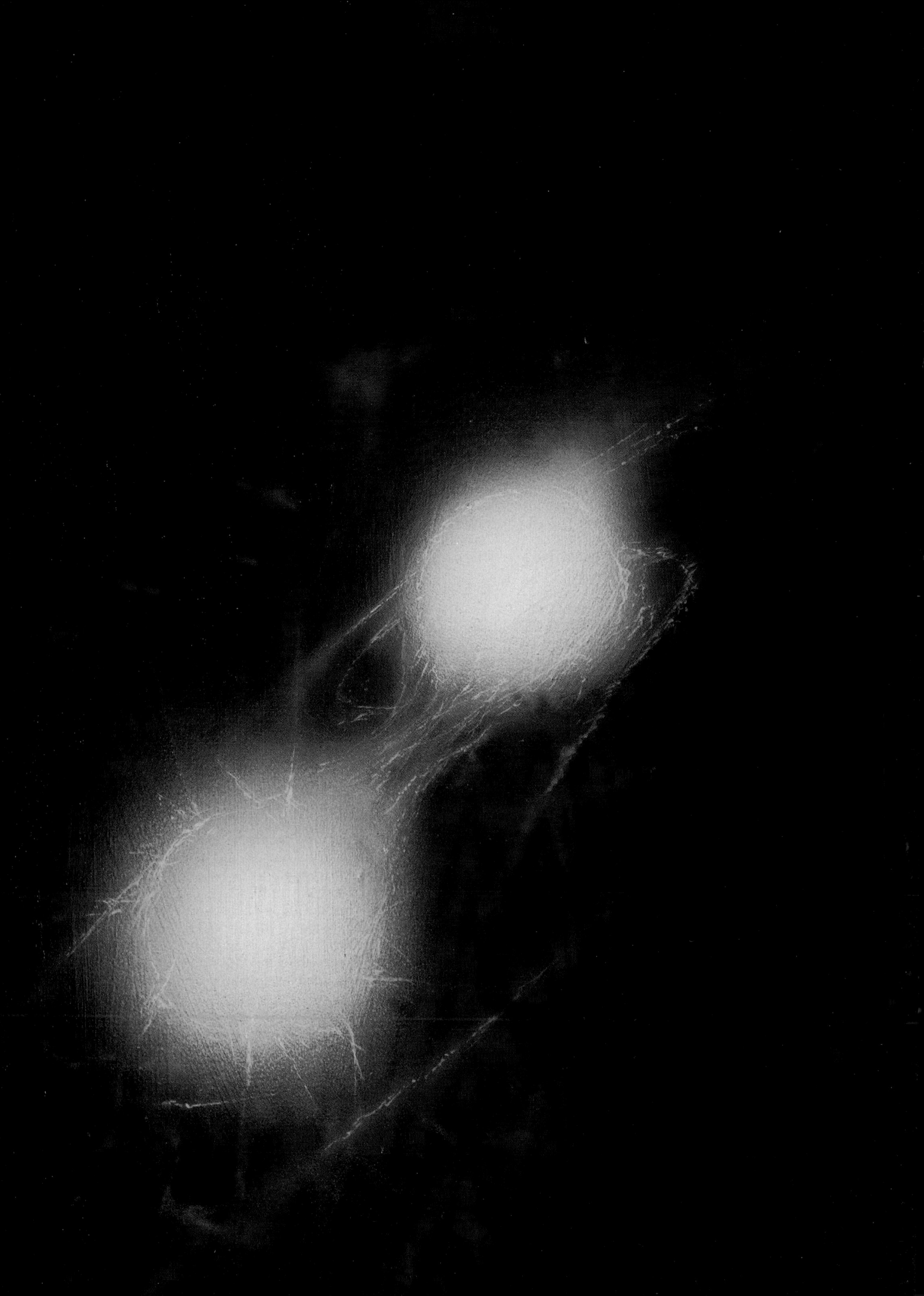

The Genesis of Binary Stars

Fast new supercomputers are giving old theories a fresh lease on life as astronomers seek to answer the thorny question, "Where do binary stars come from?"

by Alan P. Boss

DEEP INSIDE A COCOON OF DUSTY GAS, a binary star system takes shape. Recent breakthroughs have helped astronomers sketch in the origins of these stellar systems, which account for a majority of the Galaxy's stars.

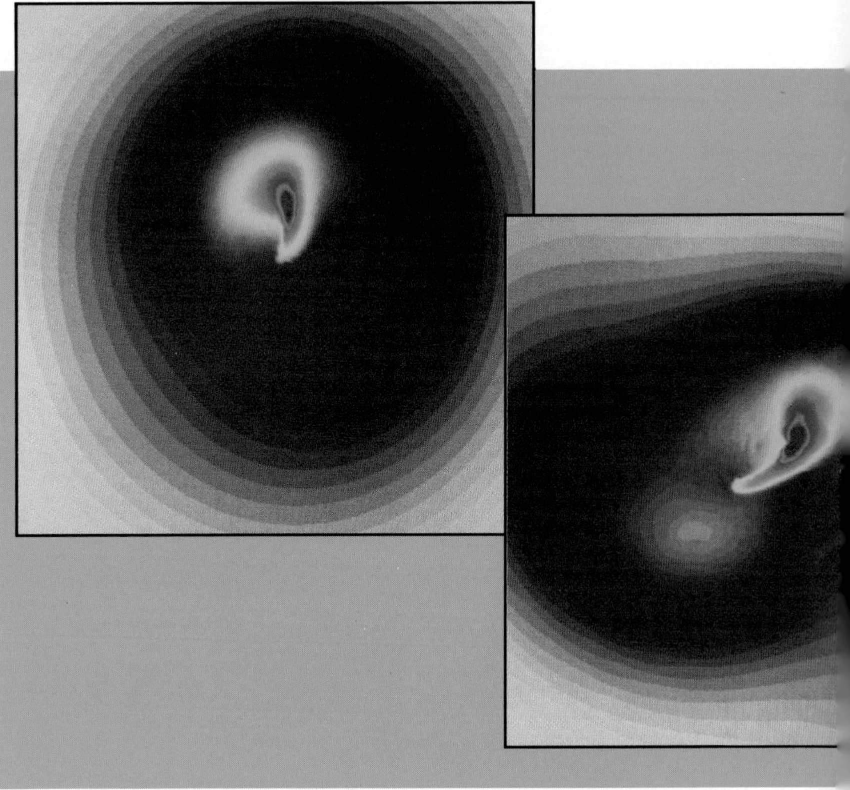

Some suns are born solitary, with no stellar siblings. Others, like the beautiful blue and gold Albireo that arches overhead tonight in the beak of Cygnus, form in twos or threes or more.

These double and multiple stars have long offered pretty sights to backyard observers, but they have been a tremendous headache to generations of astronomers. It is only in the last decade that astron-omers have made real progress in understanding how these star systems form, and this has come about mainly through the marriage of sophisticated theory to fast super-computers.

For a long time after the telescope was invented, no astronomer knew that double stars even existed. Observers simply assumed that whenever they saw two stars close together in the eyepiece it was just a chance alignment of objects at different distances from Earth.

The discovery of true binary stars — two stars actually locked in a gravitational embrace — came out of the research of William and Caroline Herschel, a brother and sister "binary" who might well be the patron saints of backyard astronomy.

The Herschels were tireless observers who regularly scanned the heavens with 6-inch to 18-inch telescopes. In 1782, they began compiling a list of double stars that drifted through their eyepieces. The list grew and grew until it soon contained so many pairs of stars that it was highly unlikely they were all the result of chance alignments. Finally, having observed three separate pairs of stars for several decades, William Herschel proved that each pair was revolving about a common central point. When his same test was applied to other pairs on the list, some of them indeed turned out to be "optical" pairs — two stars that really were unassociated. But the bulk of the doubles passed the test with flying colors: they were true binary systems.

Dominating the Neighborhood

Why do astronomers think double and multiple stars are important? Simple — nature has made so many of them. Of the ordinary main-sequence stars in the disk of the Milky Way, most are members of binary or multiple systems, rather than single stars like our own Sun.

Recent work by two astronomers has given us more information about the makeup of those systems. A. Duquennoy and M. Mayor of the Geneva Observatory have studied all main-sequence stars similar to our Sun within 75 light-years. They learned that, in fact, most of these stars were not loners, but binaries, with companion stars of varied mass. In 65 percent of the pairs, the companion star was at least one tenth as massive as the primary star. (A primary star is the more luminous one in a pair or a system.) Another 18 percent were stars whose companion may have been of lesser mass.

Duquennoy and Mayor also found numerous multiple stars in their sample, both triple and quadruple systems. These multiple stars fall into either of two basic types: hierarchical and Trapezium. Most are "hierarchical": the system consists of two stars making a close binary pair, with a companion (which may be another tight binary pair) orbiting at a distance. Alpha Centauri, with its close A-B pair and the much more distant dwarf star, Proxima, falls into this category.

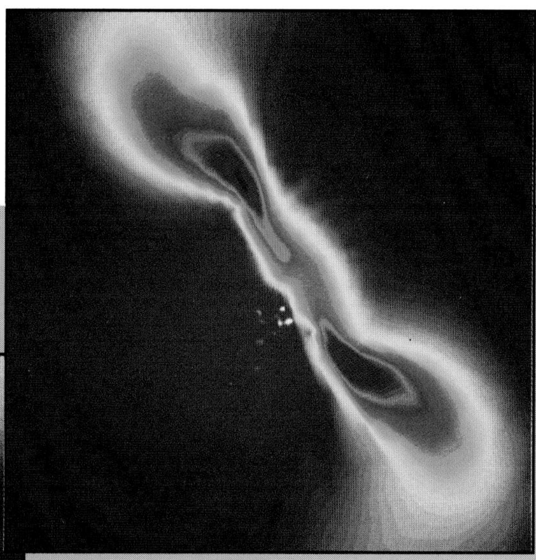

RECIPE FOR A BINARY STAR: COOL THE CLOUD THAT GIVES IT BIRTH

A cold, dusty cloud of gas is the starting point for astronomers when they try to computer-model how binary stars form. The cloud in this model is as massive as the Sun. It has a diameter of 550 astronomical units and is color-coded to show variations in density (blue is most dense, red is least).

The sequence of three images starting at far left shows that as astronomers lower the temperature in the collapsing model cloud, the outcome is a binary star rather than a single star. Notice the twin blue "cores" in the last image.

The second type, the "Trapezium" kind, is far less common. Named for the bright multiple star in the heart of the Orion Nebula, stars of this category typically have three or more stars orbiting together with roughly equal separations. Trapezium systems are rare because such a situation is highly unstable — eventually three of the stars will have a close encounter and one or two of them will be ejected. Astronomers usually find Trapezium systems in young star-forming regions like Orion, where the newborn systems have not yet had time to disintegrate. In contrast, hierarchical systems are dynamically stable and hence quite long-lived.

To a New Neighborhood

The survey of Duquennoy and Mayor looked only at mature stars. To probe the source of binaries, astronomers need to push back the age limits and examine much younger objects — the pre-main-sequence stars that have not yet fired up their thermonuclear furnaces. Robert Mathieu of the University of Wisconsin and his collaborators have found a number of spectroscopic binaries among the pre-main-sequence stars.

A spectroscopic binary is a binary system that lies far enough away that it appears in the eyepiece as a single star. But when this star is studied with a spectrograph, the lines in its spectrum periodically change position or split in two. This occurs when the orbital motion of the unseen companion star imparts a back-and-forth Doppler shift to the primary's spectral lines.

Astronomers have found such hidden binaries other ways too. Michal Simon of SUNY Stony Brook and Wen-Ping Chen of the Carnegie Institution of Washington have used lunar occultations to detect unseen companion stars in binaries. When the airless rocky limb of the Moon glides in front of an unresolved binary, it slices into the star's beam of light like the keenest of knife-edges. If the star is double and favorably aligned, one of its components will snap out of sight a fraction of an instant before the other. With an instrument that responds quickly enough, astronomers can see a clear pause in the signal as the light from the star drops to zero in two steps.

While Simon and Chen have observed relatively few stars as yet, their results show that binary systems occur among pre-main-sequence stars just as often as among the main sequence stars. This confirms the overall statistical frequency of binaries, but it also indicates that most or all binary stars are *born* that way, not made subsequently.

In pushing back toward younger and younger stars, astronomers run into an observational problem. The youngest protostars — stars still in the process of forming from their parent interstellar clouds — are swaddled in dense cocoons of dusty gas and are difficult to view. Astronomers expect, however, that new millimeter-wave radio telescopes — working at wavelengths that pass through the dust relatively unhindered — will soon "see" into such clouds and detect binary protostars. One such protobinary has already been found in the Rho Ophiuchi cloud by Alwyn Wootten of the National Radio Astronomy Observatory. The components in Wootten's protobinary orbit about 750 astronomical units apart, a typical value for binary stars, whose separations range from a few stellar radii to about 20,000 AU. (Earth orbits 1 AU from the Sun, and Pluto's average distance is 40 AUs.) Finding many more binary protostars would imply that binary systems may be created during star formation itself — the mysterious and imperfectly understood process that occurs when interstellar gas clouds collapse under gravity.

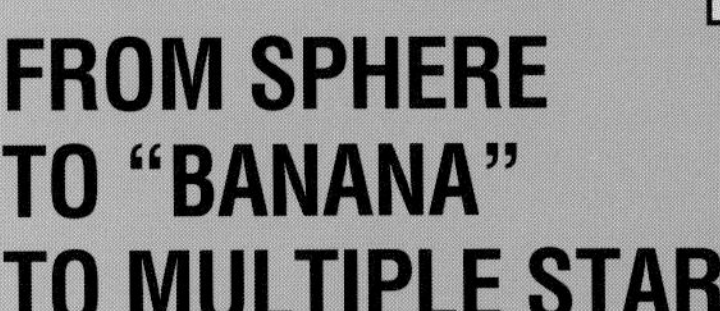

FROM SPHERE TO "BANANA" TO MULTIPLE STAR

Starting with a spherical cloud of cold, dusty gas (top), this model star system evolves into a multiple star. The diameter of the 1 solar mass cloud at the beginning is some 400 astronomical units, but this shrinks to about 250 by the last frame. (Blue denotes areas of highest density, red the lowest.)

The cloud first develops a banana-shaped region of high density as random motions cause the material to coalesce. Then other dense areas arise. The end result is a small group that will form a multiple star with three components. It's also possible that the large lump at about 8 o'clock in the bottom image would fragment in two.

Computer models show that multiple stars often take the form of a close binary pair and a distant companion. (Alpha Centauri, the star nearest to the Sun, has just this configuration.)

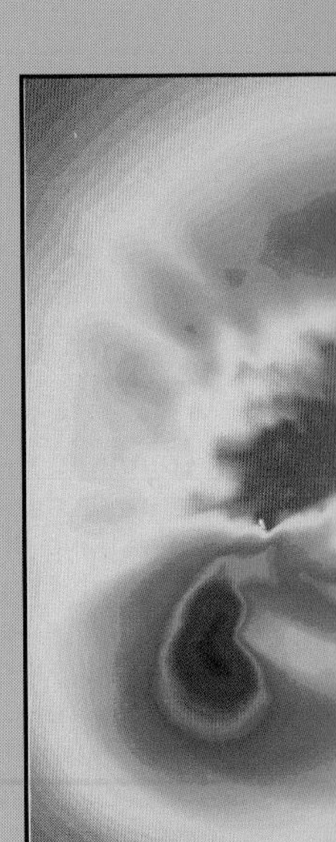

What is the Orbital Period for a Theory?

Reduced to its essence, understanding how binary stars form means learning how an interstellar gas cloud, a protostar, or a pre-main-sequence star can divide into two or more separate, self-gravitating bodies. We have seen that observations of very young binary stars are quite difficult at present because of the closeness of the objects to each other, the distance to the nearest star-forming regions, and the presence of intervening gas and dust that usually shield the process from view. Hence much of our understanding of binary star formation has come from theoretical work, particularly computer simulations.

As astronomers today juggle their complicated equations with supercomputers, they look back in wonderment at past efforts, which were often carried out with no computational means more sophisticated than pencil, paper, and logarithm tables. Theories for the formation of binary stars began almost as soon as the stars were recognized. Soon after the Herschels began finding large numbers of them, the French astronomer and mathematician Pierre Simon de Laplace suggested a separate-nuclei hypothesis for their origin. Laplace thought that binary stars might occur simply because their progenitors happened to be located close together.

Unfortunately, Laplace described his hypothesis so vaguely that it was largely ignored for two centuries. During the 1980s, however, the discovery of young stars and dense clumps of matter buried in interstellar clouds has resulted in a rebirth of Laplace's hypothesis, albeit in a different guise.

James Pringle of Cambridge University has suggested that these clumps should occasionally undergo

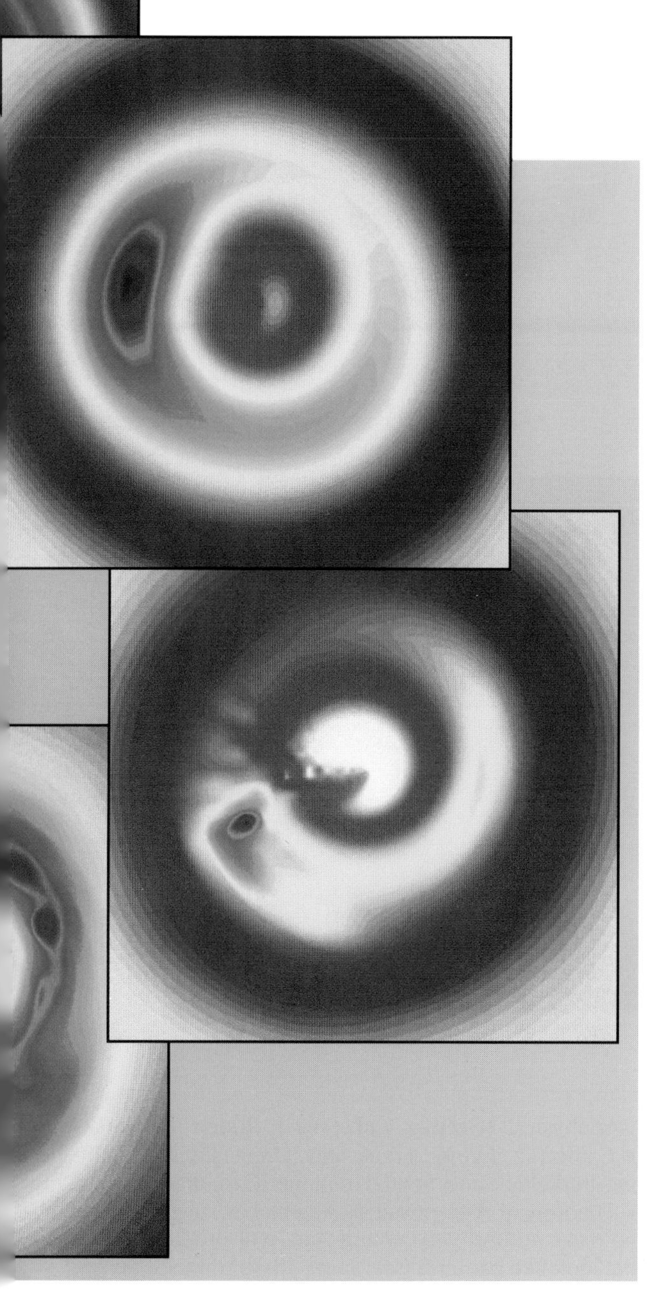

collisions with each other. The collisions should produce strongly compressed regions where the clouds collide, and these regions might then begin to collapse and lead to the formation of a binary system. Pringle's hypothesis has not yet been studied in detail, but the computer modeling of John Lattanzio at the Lawrence Livermore Laboratory and his collaborators appears to cast doubt upon it. Their work suggests that the outcome of cloud-clump collisions is more likely to be a roughly spherical cloud that may not collapse to form a binary system.

Another classical hypothesis for binary origin is capture, which posits that pre-existing single stars become trapped in a binary system. But captures are tricky. They can occur only if some mechanism exists to remove kinetic energy from the individual stars in the encounter. Otherwise the stars zip past each other on hairpin trajectories and fly apart like some kind of cosmic pinball game.

Theoreticians have proposed several mechanisms for dissipating kinetic energy. One is the three-body encounter, in which a third star is ejected and carries off the excess kinetic energy. Another is tidal dissipation, by which tides raised in each star during a close encounter rob energy from both. However, when astronomers calculate the probability of having the bodies come just close enough in exactly the right way, it turns out capture is an enormously improbable means of making binary stars — even in dense star clusters where close interactions will occur much more frequently than among the stars of the galactic disk.

Richard Larson of Yale University recently tried to revive the capture method. He pointed out that the friction experienced by a star passing through the gaseous disk of a protostar might soak up enough energy to trap the star and protostar into a binary orbit. While the frictional forces in this situation remain to be determined, astronomers think it's still unlikely that capture by this mechanism could happen frequently enough to be important — although an exception might perhaps occur in stellar clusters that are even denser than the Orion Nebula.

The final classical method is fission, long a favorite of scientists and mathematicians such as William Thomson (Lord Kelvin), George Darwin (son of Charles Darwin), James Jeans, and Henri Poincaré. The fission hypothesis, which had its heyday in the late 19th and early 20th centuries, is based on the notion that if a body spins fast enough it will fly apart. Scientists worked out the precise speeds of rotation at which bodies of uniform density would become unstable, but they were unsure exactly what form the instability would take. They hypothesized that because the instability begins with a tendency of the body to divide in two, this might very well continue and lead to the formation of a binary star.

The fission hypothesis may not be new, but it has attractive features. As a pre-main-sequence star contracts toward the main sequence, it will spin faster and faster, like a whirling skater who draws

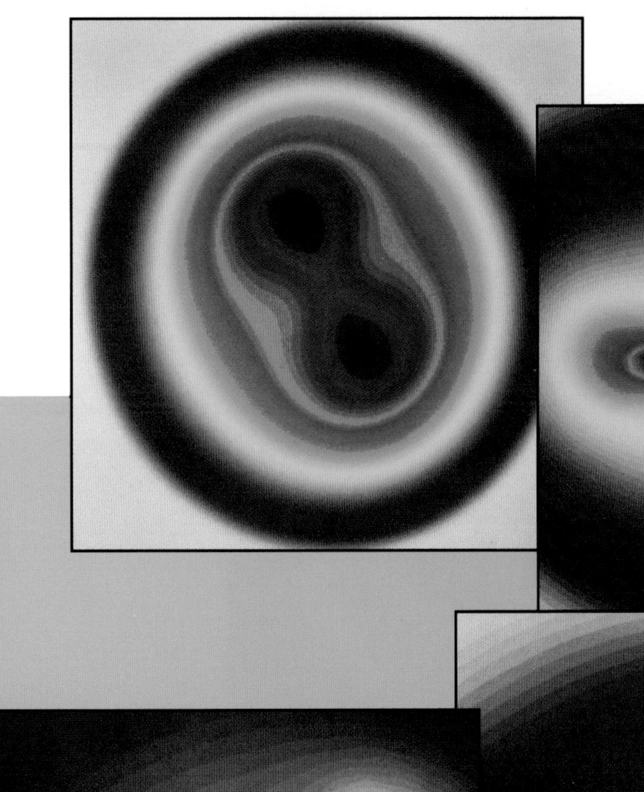

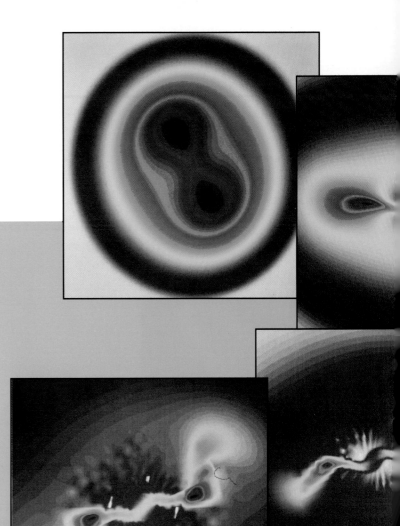

PAIRS WITHIN PAIRS WITHIN PAIRS...

If the cold cloud of gas and dust has an elongated shape (as shown here at top center), the result often leads initially to a binary pair of stars, according to the computer models.

This cloud, which weighs as much as the Sun and has an initial diameter of more than 4,000 astronomical units, will first form a binary pair about 1,000 astronomical units apart. But both lumps are probably large enough to fragment again, which will create a binary hierarchy, with four stars grouped in two pairs of two (near right).

This kind of arrangement, gravitationally stable, can be found throughout the sky.

in her arms. If the protostar should spin fast enough, fission instability could develop and the star break in two.

The hypothesis looks good enough that over the last decade astronomers spent much computer time modeling plausible outcomes for fission instabilities. Unfortunately, the results all show that fission does *not* lead to binary star formation. Instead, spiral arms form similar to those in spiral galaxies. As these expand, they remove angular momentum from the center of the body, lowering its rotation rate below the critical value for instability. The spiral arms then coalesce into a relatively low-mass disk in the equatorial plane of the body. Observational evidence backs up the calculations, because young protostars are known to have such disks. In fact, Earth and the other planets formed from just such a disk around the newborn Sun.

Another hypothesis has been put forward recently by Fred Adams of the Harvard-Smithsonian Center for Astrophysics. Together with Steven Ruden (University of California, Irvine), Frank Shu (University of California, Berkeley), and Scott Tremaine (Canadian Institute of Theoretical Astrophysics), Adams has suggested that the disks around very young stars will produce waves that look like bananas (see pages 38-39) rather than spirals. These bananas, they suggest, might eventually break up and form a second protostar in addition to the one already at the center. The result: a binary system.

This mechanism has not yet been investigated in sufficient detail to know if it works. But once again, this kind of instability shares similarities with fission instability and these hint at a similar — and equally negative — outcome.

Big Binary Star Machines

Over the last decade or so, I and many other astronomers worldwide have been developing computer programs — "codes," in the jargon — to study how

gas clouds can collapse and form stars. The codes involve the numerical solution of mathematical equations that determine the flow of gas, dust, and radiation in a cloud. The equations are similar to those used to predict weather on Earth. However, the astrophysical codes must also solve for the cloud's self-gravity, because it is gravity that ultimately holds each binary object together.

These programs are terrifically complex — they have to be if we want to model even modestly a 3-dimensional gas cloud light-years across — and they require large amounts of computer time on the fastest supercomputers available. (It's no accident that the research in the field has gone hand-in-hand with the development of super number-crunchers.) In addition to the computer code I have developed, many other similar codes have been developed independently by other astronomers.

The existence of independently written codes might look like a wasteful duplication of effort, but it confers huge benefits by permitting us to calculate the same theoretical model with each of our codes and to compare the results. In part because the codes are complex, we can't ever be completely certain how valid our results are, but comparisons let us sort out the real physics from the spurious. It's reassuring, however, to note a satisfying amount of agreement, including both codes that treat the cloud as a continuous fluid and codes that model the cloud as a collection of smoothed particles.

Today's most promising approach to the problem of how to make binary and multiple stars seems to be fragmentation. This refers to the possibility of an interstellar cloud's breaking up into two or more protostars during the time the cloud is collapsing. Almost forty years ago Fred Hoyle pointed out that as a gas and dust cloud collapses and becomes denser, objects of ever-smaller mass might be able to fragment out of it. The computer codes have been instrumental in discovering under what conditions a collapsing interstellar cloud is likely to produce a binary system.

Generally speaking, nature can make a binary star when the gas cloud is cold, its center is moderately condensed, it has appreciable rotation, and it is not controlled by magnetic fields. Unfortunately, observers have not yet been able to identify interstellar clouds that are now on the verge (or in the early phases) of collapse. Such observations will be critical for determining whether or not real clouds actually do begin to collapse from the configurations that fragmentation theory says are likely to lead to the formation of binary stars.

The illustrations show some of the possible outcomes for the collapse of interstellar clouds. When the initial cloud is strongly distorted and cigar-shaped — we have seen some of these in nature — the cloud tends to collapse and form a binary protostar. As each of these protostars undergoes its own collapse, a second phase of fragmentation may set in, forming a binary system on a smaller scale. This yields a hierarchical multiple system. But when the cloud is not as strongly distorted to begin with, it may collapse to form a ring-like structure that eventually fragments into a small number of protostars with varied masses. In this case, a Trapezium-style multiple system may result.

However successful, fragmentation is unlikely to be the last word in binary formation theory. But astronomers are finding it useful because it appears capable of explaining many kinds of binary and multiple stars as well as key properties of them, such as their tendency to have highly eccentric (non-circular) orbits and unequal masses.

A number of new computer codes have been developed in the last few years and are now being applied to the study of various problems associated with binary star formation. This is clear evidence for a resurgent interest in this difficult and abstruse field. New theory is being coupled with strong ongoing observation efforts aimed at detecting and describing young binary systems. We can look forward to seeing great progress in understanding these beautiful and intriguing stellar systems. □

Alan P. Boss is an astrophysicist at the Department of Terrestrial Magnetism of the Carnegie Institution of Washington, Washington, D.C.

STARS TANGLE AND EXCHANGE MATES deep in the heart of the Orion Nebula. Some 80,000 years ago a binary star passed into the Orion complex at high velocity, gravitationally interacting with another star. One star escaped and the bystander began to orbit the original primary. Illustration by Joe Bergeron.

ASTROPHYSICAL RESEARCH

Encounter in ORION

A young double star fleeing the Orion Nebula at high speed provides insight into how binary stars form — and how they escape their birthplaces.

by Ken Croswell

Stars, like people, usually have mates. Though our lonely Sun is single, most stars in the Galaxy — such as Sirius and Procyon — have partners to keep them company through the night.

People, of course, go through an elaborate process during which they meet and court their mate. How double stars form, however, remains unknown. Some stars may be born double, while others, like people, may start their lives single and encounter their lifelong companion later.

"There's no answer yet to this question," says Robert Mathieu of the University of Wisconsin at Madison, who studies binary stars shortly after they are born. "It's the hot subject right now in star formation theory."

To learn how binaries form, Mathieu must catch the stars while they are still very young — while the stars are in the pre-main-sequence phase. During the pre-main-sequence phase, the star gradually contracts, generating light by converting gravitational energy into heat.

As the pre-main-sequence star gets smaller, its core gets hotter and hotter, until the star ignites its hydrogen and joins the main sequence. On the main sequence, a star generates energy by fusing hydrogen into helium at its core. Ninety percent of all stars, including the Sun, are main-sequence stars.

Pre-main-sequence stars, however, are rare, because a star spends only a brief amount of time in this stage of evolution. Therefore, binaries containing two pre-main-sequence stars hold vital clues to how binaries form.

In some cases, stars might be just like people, mating long after they are born.

Recently, one such pre-main-sequence binary in the Orion Nebula has attracted a lot of attention: Parenago 1540, first catalogued by the Russian astronomer P. P. Parenago in 1954. Parenago 1540 is a remarkable double star which implies that, at least in some cases, stars are like people in that they form unions well after their birth.

When Mathieu and his colleague Laurence Marschall of Gettysburg College measured the age of Parenago 1540, they found that one star in the binary appears to be twice as old as the other, suggesting that the two did not form together. Furthermore, Parenago 1540 is escaping the Orion Nebula at high speed. This indicates that the binary has suffered an encounter with another star that has tossed Parenago 1540 out of the star-forming region of the Orion Nebula for good. This may explain how the stars joined together.

"If we can decipher Parenago 1540," says Mathieu, "we may gain unique insight not only into the formation of binaries but also into the dynamics of young star clusters."

Young Couples

Like all binaries, pre-main-sequence binaries come in two main types, visual and spectroscopic. Astronomers have known about visual pre-main-sequence binaries since the 1940s. In these systems, astronomers can see the two different stars.

Unfortunately, the nearest pre-main-sequence stars are 500 light-years away. Hence, to appear as a visual binary the two stars must be separated by huge gaps of space. Because of this large separation, the stars move about each other slowly, taking thousands of years to complete an orbit. This makes it impossible for astronomers to determine the stars' orbits and masses.

During the last ten years, however, astronomers have discovered a new type of pre-main-sequence binary, the so-called double-lined spectroscopic binary, which allows astronomers to derive orbits and masses for the two stars.

Two stars in a pre-main-sequence double-lined spectroscopic binary lie so close together that they typically revolve around each other every few days or weeks. Because the stars lie close together, astronomers cannot see the two stars individually.

Instead, spectra reveal two different sets of spectral lines, one set for each star. As the stars orbit each other, one star approaches us as the other recedes from us. This orbital motion produces periodic redshifts and blueshifts of the stars' spectra that reveal the binary's orbital period.

The spectra also measure how much more mass one star has than the other, because the lighter star orbits faster than the heavier one. Mass is crucial, because mass dictates a star's rate of evolution and is therefore a prerequisite to measuring the star's age. The more mass a star has, the faster it evolves. Knowing both a star's mass and its state of evolution allows us to deduce its age. If two stars in a binary have equal ages, they probably formed together. If the two stars have different ages, then more interesting possibilities arise — as may be the case for Parenago 1540.

A Date with Parenago 1540

Pre-main-sequence double-lined spectroscopic binaries therefore hold the key to determining the ages of both stars in a young binary. Astronomers found the first such system, V826 Tauri, in 1982. Since then, only a few more have been discovered. One of the most tantalizing is Parenago 1540.

Mathieu and Marschall first studied Parenago 1540 during the mid-1980s, as they were observing stars in the Trapezium star cluster. The Trapezium cluster lies at the center of the Orion Nebula and contains the bright multiple star Theta Orionis. The Trapezium gets its name because the four brightest stars of the Theta Orionis system make the shape of, you guessed it, a

PARENAGO 1540, visible in amateur telescopes, can be seen in this photograph as a faint star near the outer edge of the Orion Nebula. Photograph by Mike Sisk.

trapezium. But the cluster contains hundreds of other young stars as well. Parenago 1540 is one of them, shining at magnitude 11.33 and lying just 10 arcminutes west of Theta Orionis. Like its mates, Parenago 1540 is 1,500 light-years from the Sun.

Mathieu and Marschall observed Parenago 1540 as part of a project they conducted to measure the velocities of the Trapezium cluster's many stars. To do so, they acquired spectra that revealed the redshifts or blueshifts of the stars in the cluster.

"Parenago 1540 was double-lined," says Mathieu, "so right away it intrigued us. It was only the third pre-main-sequence double-lined spectroscopic binary found, so its very existence was an interesting thing in itself." From the periodic shift of the spectral lines, Mathieu and Marschall determined that the two stars orbit each other every 33.7 days, with the heavier star 1.3 times more massive than the other.

But the story didn't end there. "We went through an analysis and found that when we tried to place both components of the binary on the H-R diagram to get masses and ages — by using standard techniques for weighing and dating pre-main-sequence stars — the two stars did not appear to be the same age."

How can one measure a star's age? It's never easy. We know the age of the Sun — 4.6 billion years — only because we can date rocks on the Earth and Moon. For many other stars, however, determining age is impossible. The H-R diagram, a plot of luminosity versus color, helps, especially for pre-main-sequence stars. Bright stars lie near the top of the H-R diagram, dim stars toward the bottom; hot blue stars to the left, cool red stars to the right.

Parenago 1540 is going on a honeymoon — a honeymoon that may reveal how the two stars first dated.

On the H-R diagram, main-sequence stars form a diagonal line from the upper left (bright and blue) to the lower right (faint and red). Pre-main-sequence stars are bright and cool, so on the H-R diagram they lie above and to the right of the main sequence. As a pre-main-sequence star contracts, it moves toward the main sequence.

In principle, stellar evolution theory can show how a pre-main-sequence star with a given mass changes its position on the H-R diagram as the star ages. Therefore, by measuring the luminosity and temperature of a pre-main-sequence star and thereby the star's position on the H-R diagram, one can establish the star's mass and age.

In the case of Parenago 1540, Mathieu and Marschall estimated that the primary has a mass around 2.25 Suns, whereas the secondary has a mass of 1.7 Suns. At present, these stars are cool orange K-type stars. But when they reach the main sequence, both will be white A-type stars, brighter and hotter than the Sun. In fact, Parenago 1540 A will someday resemble Sirius (spectral type A1) and Parenago 1540 B will look like Altair (spectral type A7).

Of course, Parenago 1540 has not yet reached the main sequence. This gives astronomers the chance to date the two components. Two stars of the same mass and same age should also have the same luminosity and temperature — that is, they should have the same position on the H-R diagram. Parenago 1540 A and B have fairly similar luminosities. This implies that the two stars — if of equal age — should also have similar masses. But they don't.

"That's the surprise," says Mathieu. "The mass ratio for the system is relatively large, considering that both stars seem to be of roughly comparable luminosity." This result may suggest that the stars are different ages. In 1988, Mathieu and Marschall reported that their best estimates gave Parenago 1540 A an age of one million years and Parenago 1540 B an age of only half a million years.

The Problems of Dating a Younger Star

It's tricky to measure the ages of young stars, however. Taken at face value, the different ages of Parenago 1540 A and B imply that the two stars did not form together. But Mathieu is the first to warn against overinterpreting the result.

"It's a very, very difficult analysis," he says. "There are a lot of steps from the data to the result, all of which have ambiguity, as most scientific analysis does.

"I have been very careful when I refer to Parenago 1540," says Mathieu. "An attempt to reduce it to its essence can make the result appear far stronger than it is. You go through the error analysis, discuss all the qualifications, and then it often boils down to somebody saying these stars didn't form at the same time. But it's not that definitive."

Determining a pre-main-sequence star's correct placement on the H-R diagram is the first problem. In order to do that, one must know the star's luminosity and temperature. But the luminosity can be wrong because obscuring disks often surround pre-main-sequence stars. These disks, where planets possibly form, block some of the stars' light, which is then re-radiated in the infrared. Errors may also appear in the derivation of a star's temperature from its spectral type. Furthermore, dust lying between us and Parenago 1540 can affect the observations. Even worse, more dust may lie in front of one star than the other.

Problems also exist with the theory. Exactly how does a pre-main-sequence star contract toward the main sequence? No one has ever seen a star do it. So astronomers must rely on theoretical calculations of how pre-main-sequence stars evolve. The question is complicated by the fact that if disks surround the stars, those disks could affect the stars' evolution. Furthermore, says Mathieu, "The presumption that single-star pre-main-sequence evolutionary tracks are relevant for two stars that may have formed in very close proximity is not at all clear." Indeed, one star in a binary may disturb the other's evolution, rendering impossible an attempt to determine ages from standard theory that has been developed for single stars.

Despite the possible errors, Mathieu and Marschall believe that Parenago 1540 may contain two stars of differing ages. And another pre-main-se-

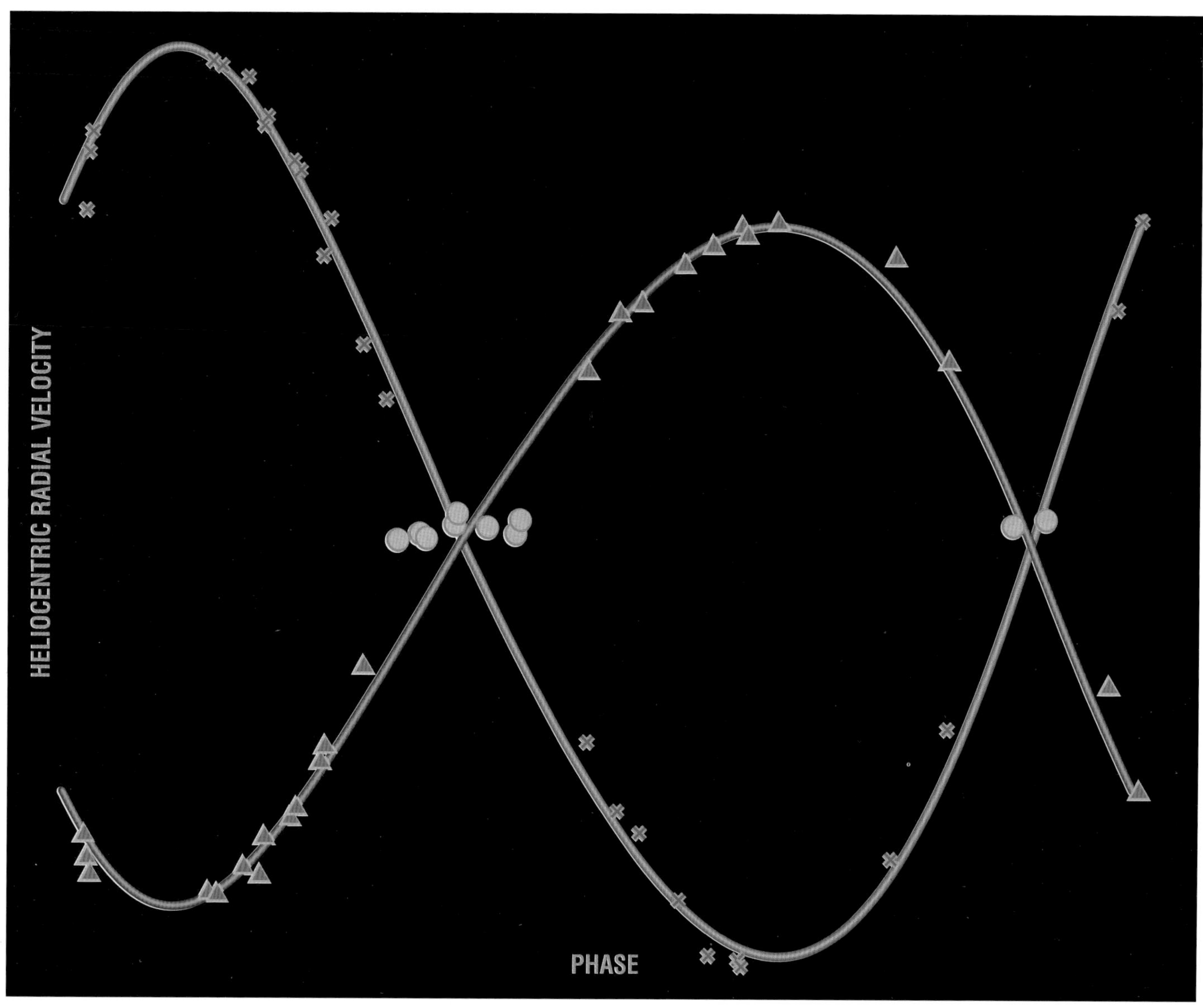

A RADIAL VELOCITY CURVE of the Parenago 1540 system shows the relative motions of the primary (marked by triangles) and the secondary (marked by circles). The more massive primary orbits more slowly than the secondary, which has several important implications.

quence spectroscopic binary — V4046 Sagittarii — has the same characteristic. V4046 Sagittarii consists of two stars with equal masses of 0.7 Suns. If the two stars are the same age, they should also have the same luminosity and temperature. Yet one star is twice as bright as the other.

"A student of mine, Chi-Wai Lee, is presently doing a dissertation on several pre-main-sequence binaries," says Mathieu. "He's developing more sophisticated analysis techniques than Larry and I used. I'm hoping that the sum of all these binaries together will provide a more definitive statement of whether or not Parenago 1540 is unusual and whether or not our result is to be believed."

In one respect, though, Parenago 1540 is definitely unusual, in a way that suggests the stars do not have equal ages: Parenago 1540 is shooting straight out of the Trapezium.

Parenago 1540's Honeymoon

Like many newlyweds, Parenago 1540 is going on a honeymoon — a honeymoon that may reveal how the two stars first dated and mated. The star system is fleeing its place of marital union at high speed, never to return. Astronomers know this because Parenago 1540's proper motion points due west, directly away from the Trapezium. By tracing back its velocity, Mathieu and Marschall estimate that the star got ejected from the cluster only 80,000 years ago.

"This may close the picture," says Mathieu. "It's conceivable that the reason the two stars are different ages — if they are — is that they did not in fact form together. Rather, the pairing may have been produced during a dynamical interaction that ejected the binary from the cluster."

The most likely scenario for Parenago 1540's ejection works like this. A single star in the Trapezium cluster encounters a binary star in the Trapezium. The single star splits the binary by kicking out one of the binary's two stars and mates with the remaining star — it's a homewrecker! The new binary then converts some of the original binary's orbital energy into kinetic energy, and the new binary flies away from the cluster at high speed.

Birth of a Double Star

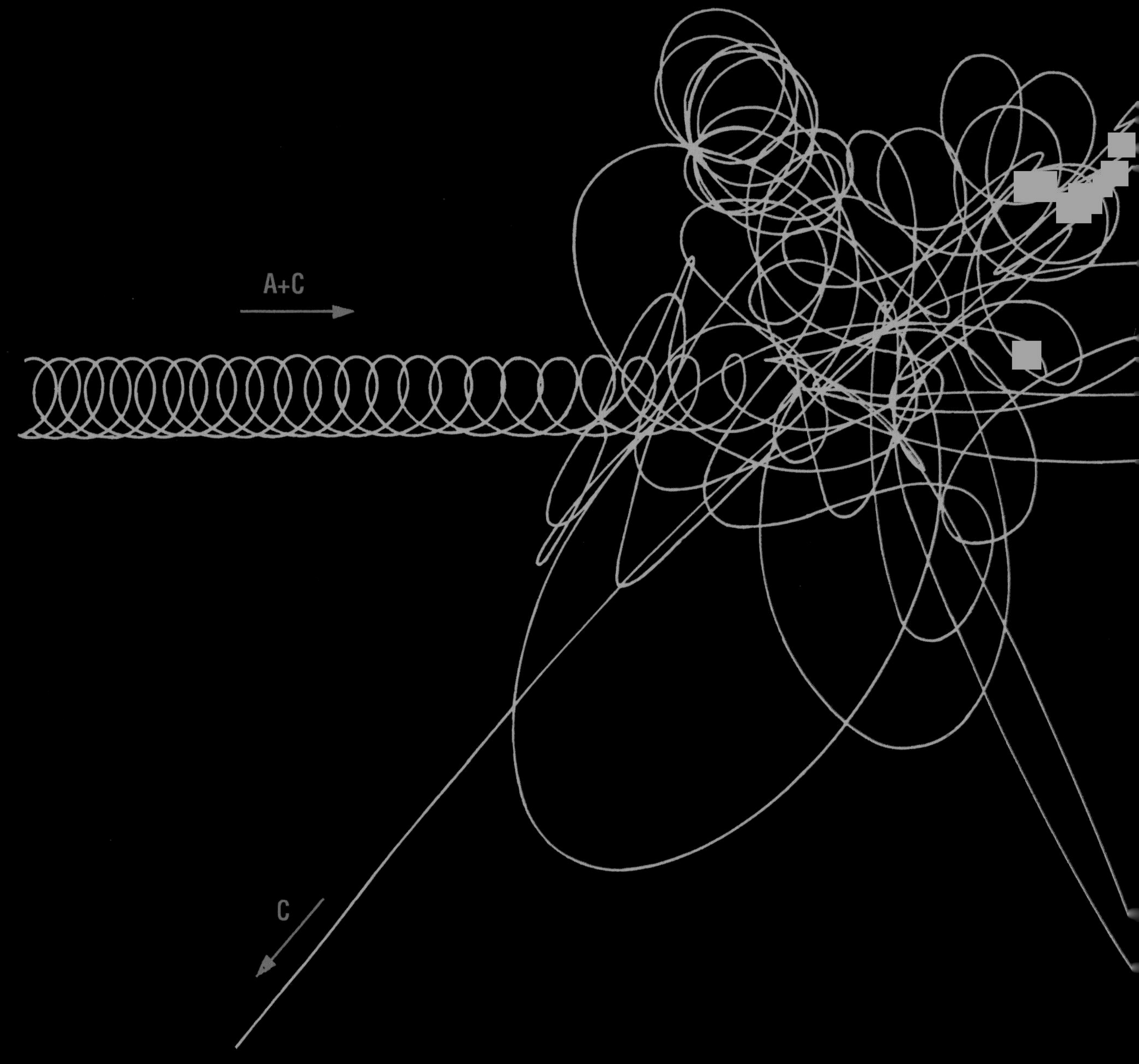

"Stellar dynamicists have been predicting for a long time that such encounters should occur in star clusters," says Mathieu. "To actually see evidence for it is very exciting."

Other scenarios for Parenago 1540's ejection also exist. Two isolated stars without proto-planetary disks cannot come together and form a binary, for the stars fly past each other and cannot capture each other. If the stars have disks, though, the disks can dissipate the stars' energy of motion so that the two stars might be able to become a binary.

Another possibility — which works even if the stars have no disks — involves three separate stars that come together simultaneously, producing a bi-

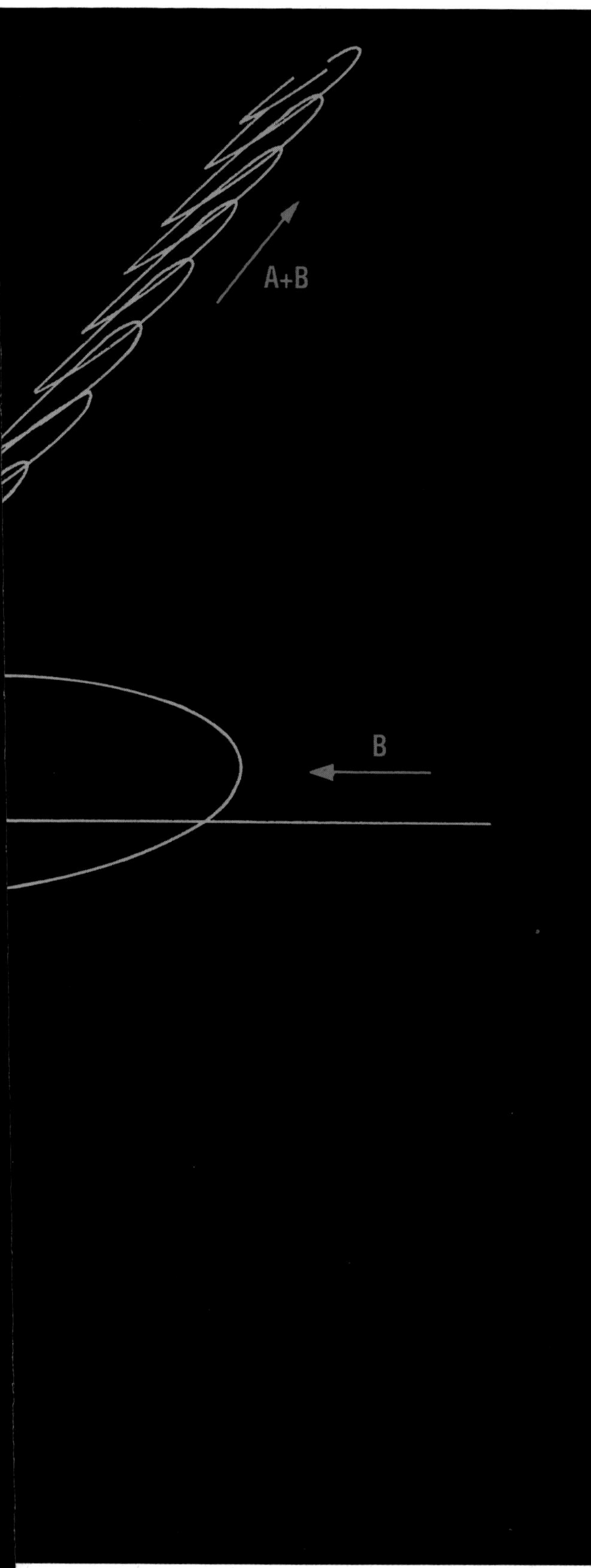

A GRAVITATIONAL DANCE occurred between Parenago 1540 and a third star. This simulation shows the original binary entering at high speed from the left and encountering the third star (at right), as well as a new binary consisting of stars 1 and 3 exiting at upper right.

nary and a single star. In this case, the single star carries away the extra energy and allows the other two to mate.

In all these scenarios, Parenago 1540 A and B form separately from each other and so should not have equal ages. If Parenago 1540 A and B actually have the same age, an encounter could still cause their flight from the cluster: a binary can meet a single star, remain intact, and still receive a shot of energy that sends the binary out of the cluster.

Intriguing though these scenarios are, one problem exists, for Parenago 1540 has a low eccentricity — its orbit is more circular rather than elliptical. "If an exchange of stars has occurred, you expect the binary that is ejected to have a very eccentric orbit," says Mathieu. But Parenago 1540's orbital eccentricity is only 0.12, close to circular.

"The trouble is, you're dealing with a case of one. Although the probability of getting an eccentricity of 0.12 is small, it's not zero — in fact, it's not even negligible. So it doesn't worry me, but obviously I'd feel better if the eccentricity were 0.8!" Furthermore, if the two stars of Parenago 1540 had dust disks around them prior to their mating, these disks could have produced a more circular orbit.

Longtime Companion

Even though its past is not yet clear, Parenago 1540 has a secure future. Both stars in the binary will contract and heat up, changing from orange to yellow to white. A few million years from now, the more massive star will ignite the hydrogen at its core and join the main sequence. Then, a few million years later, the less massive star will follow suit. What was once a pre-main-sequence binary will become a binary consisting of two full-fledged A-type main-sequence stars.

And despite their perhaps stormy past, the two stars will stay together. Both are escaping the dense Trapezium cluster at high speed; from now on, they will orbit the Galaxy as a pair. Chances are, they will never again come close enough to another star to be torn apart.

Meanwhile, Parenago 1540 might make us wonder about our own solitary Sun. If Parenago 1540 formed as a result of an exchange between a binary and a single star, creating a new binary and a new single star, the star that is now single was once part of a binary.

Might the same thing have happened to our Sun? Four and a half billion years ago, as it was being born in a star cluster, did our Sun have a faithful companion that circled it — a companion that was torn away by a bigger star? And if so, where might that lost star now be? On the other side of the Galaxy?

We may never know. But Parenago 1540 should at least raise the possibility that, during its youth, even our single Sun may have felt the attraction of a close stellar companion. □

Ken Croswell is an astronomer in Berkeley, California, and a frequent contributor to ASTRONOMY. He also writes for New Scientist, Science, *and* Star Date. *Ken's last ASTRONOMY article was "Will the Lion Roar Again?" in the November 1991 issue.*

Regal Rigel

Rigel is far more than the brightest star in Orion. It's also one of the biggest, hottest, and most luminous stars in the Galaxy.

by Peter Jedicke and David H. Levy

No where in the heavens is there a constellation more recognizable than the mighty warrior Orion. And with seven stars shining at 2nd magnitude or brighter, Orion features more bright stars than any other constellation. The Hunter's eastern shoulder, Betelgeuse, contrasts with the rest because of its ruddy hue. The other six stars all shine with a brilliant diamond-blue color: the western shoulder Bellatrix; the belt stars Alnilam, Alnitak, and Mintaka; and the two knees, Saiph and the brightest of them all, Rigel.

Stars like Rigel and the other bright blue stars in Orion dominate the night because they represent the sky's best compromise between rarity and superlative. Less than one star in a thousand is as hot and luminous as Rigel, but they are easily visible across a thousand light-years of space. Although there are marginally brighter stars, they are too few to look impressive as a group. And cooler stars, of which there are legion, literally pale in comparison. You won't find a single Rigel-type star among the three hundred stars nearest to Earth, but there are a dozen or more among the three hundred brightest.

Nature may have been reluctant to encourage the formation of Rigel and his brethren because they require so much more mass than the Sun or dimmer stars. Rigel is a huge star: as much as 25 solar masses contained in a sphere that stretches across 65 times the Sun's diameter. In our Sun's place, Rigel would almost fill the orbit of Mercury. Although this is not nearly as large as the bloated orbs of the red supergiants like Betelgeuse, Rigel is far denser, hotter, and more spendthrifty with its nuclear fuel reserves. There's a lot more excitement among the hot blue supergiants than among the cool reds.

Rigel and its neighbors appear blue-white because their surfaces are extremely hot. A star's color is a direct result of how hot its surface is, with hotter stars emitting more of their light at shorter (bluer) wavelengths. The temperature at the surface of Rigel is nearly 10,000 kelvins, which is about twice as hot as the Sun's yellow surface and three times the surface temperature of ruddy Betelgeuse.

Rigel's high temperature implies tremendous brightness as well. Each square inch of Rigel's surface radiates some fifteen times the energy of each square inch of the Sun's surface and nearly one hundred times the energy of each square inch of Betelgeuse. This profligate rate of energy generation makes Rigel bright, but it also raises an interesting question — why doesn't Rigel appear even brighter?

A Star of Superlatives

The brightness of any star depends on two factors: its luminosity and its distance. The Sun is so bright simply because it is so close — if it were as far away as some of the nearer stars it would appear unimpressive. Sirius, the brightest star in the sky, is somewhat more luminous than the Sun but is bright largely because it lies only nine light-years from Earth. Rigel and the other blue stars in Orion are all a thousand — and some more than two thousand — light-years away, lonely lanterns shining down a long cosmic tunnel. Even Betelgeuse, at 650 light-years' distance vastly farther than Sirius, is a heavenly neighbor compared to Rigel and the blue stars of Orion.

Astronomers determine a star's intrinsic brightness from its spectrum. By scrutinizing thousands of stellar spectra over the past century, astronomers have categorized stars into various types, each of which have a distinct luminosity. Rigel belongs to spectral class B8, which means it shows features characteristic of neutral hydrogen. Further, the sharpness of these spectral features indicates that Rigel is a brilliant supergiant of what is called luminosity class Ia. These two ways of categorizing stars give astronomers a remarkably reliable measure of the luminosity of any star, even if its distance is not known beforehand.

For Rigel, its spectrum reveals a star whose luminosity is some 50,000 times that of the Sun. That means if Rigel were only 10 parsecs (32.6 light-years) from Earth — a standard distance astronomers use to compare stellar brightnesses — it would shine at magnitude –7.1. At the same distance, the Sun would shine weakly at magnitude 4.8 and appear as a faint naked-eye star.

Not only is Rigel different from the Sun on the outside, but the two stars are different beneath the surface as well. Rigel's tremendous luminosity originates in the star's interior, where the temperature reaches an incredible 100 million kelvins. (In contrast, the Sun's internal temperature is a rather modest 15 million kelvins.)

This temperature difference causes Rigel to be

RIGEL BLASTS the tortured surface of a nearby planet with intense radiation. In the distance, two companion stars shine brightly. All illustrations by MariLynn Flynn.

powered by a different set of nuclear reactions than is the Sun. Whereas most of the Sun's energy comes from protons fusing together to form helium via the proton-proton chain, Rigel's superhot interior generates much more energy by fusing three helium nuclei into one carbon nucleus in the triple-alpha process. The Sun is just not hot enough to take advantage of the more powerful reactions that fuel Rigel's high luminosity.

Rigel's interior differs from the Sun's in one other fundamental way. The temperature drops much faster as you move away from the center of Rigel than it does as you move out from the center of the Sun. This temperature drop is so steep in the inner parts of Rigel that energy is transported by moving currents of hot gas. Called convection, this process carries heat out of Rigel's interior very efficiently. Farther from Rigel's center the temperature changes more slowly, and energy is transported by electromagnetic radiation instead. In the Sun, on the other hand, the situation is reversed — energy is transported by radiation in the interior while convection dominates in the outer layers.

More than Meets the Eye

When we observe Rigel with a telescope, the hot blue supergiant is not all that greets us. Rigel is a system of at least three stars whose nature has fooled some of the great observers. The intriguing story of Rigel's multiplicity began in the 1830s, when double-star observer F. G. W. Struve first saw a companion to Rigel. The companion, designated Rigel B, appeared 9.5 arcseconds away from Rigel itself, which corresponds to a separation of about 2,500 AU at Rigel's distance. The companion has the same B8 spectral type as the primary star Rigel A, but is a main-sequence star and thus much smaller and fainter than Rigel.

RIGEL: A STAR OF SUPERLATIVES
Sun
Mass (Sun = 1) 1.0
Diameter (miles).......... 865,000
Diameter (Sun = 1)............... 1.0
Apparent Magnitude......... –26.8
Absolute Magnitude............ 4.8
Luminosity (Sun = 1)............ 1.0
Spectral Type .... G2 V
Surface Temperature ..... 4800 K
Color.............................. Yellow
Rigel A
Mass (Sun = 1)............................ 25
Diameter (miles)............ 56 million
Diameter (Sun = 1)...................... 65
Apparent Magnitude.................. 0.1
Absolute Magnitude................ –7.1
Luminosity (Sun = 1)........... 50,000
Spectral Type .... B8 Ia
Surface Temperature .... 9500 K
Color.............................. Blue-white

Since Struve's original measurements, however, no one has reliably observed any change in the relative positions of Rigel A and B. Nevertheless, the two

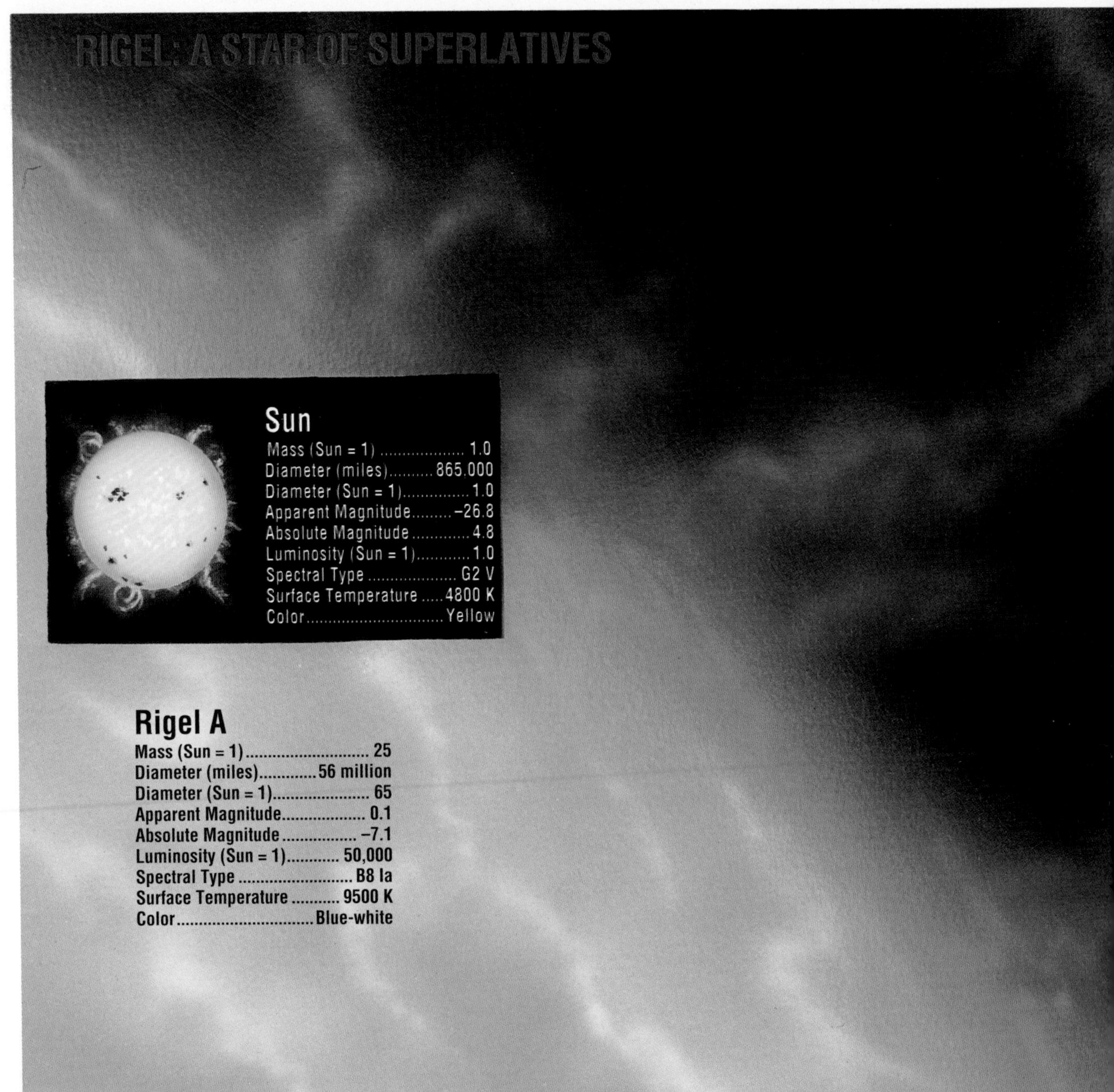

stars have nearly identical proper motions — meaning their apparent motions across the sky are the same — and velocities along our line of sight. It is highly unlikely that their common motion through space is a coincidence, so astronomers presume that the pair is physically related and that Rigel A and B do orbit each other, albeit very slowly.

The Rigel system apparently gained another member in 1908, when the Canadian astronomer John Stanley Plaskett reported that Rigel A was itself a spectroscopic binary. He inferred this from spectra of Rigel that seemed to show regular variations in the star's radial velocity. Radial velocity is the speed of an object in the direction either toward or away from an observer. When a star moves along our line of sight, the wavelength of the light we receive from it is shifted by an amount proportional to the star's radial velocity. If the star is moving toward us, the light is shifted toward shorter wavelengths, and when the star moves away from us the light shifts to longer wavelengths.

A star in orbit around a companion star will seem to move alternately toward and away from Earth as it circles the companion. The resulting back-and-forth shifts in the spectral features of the primary star betray the binary system, even if the companion star is invisible. This is what Plaskett saw in the spectrum of Rigel A. Because he could not see any spectral features from the companion star, he assumed the companion was too faint to be captured by his spectrograph.

Despite his painstaking observations, however, it appears that Plaskett was wrong. Decades later, improved techniques still have not revealed the spectral lines of Rigel A's putative companion. Astronomers now suspect that the observed variations in radial velocity are caused by pulsations of the outer atmosphere of Rigel A.

Meanwhile, the visible companion — Rigel B — played the lead role in a subplot that unfolded rather

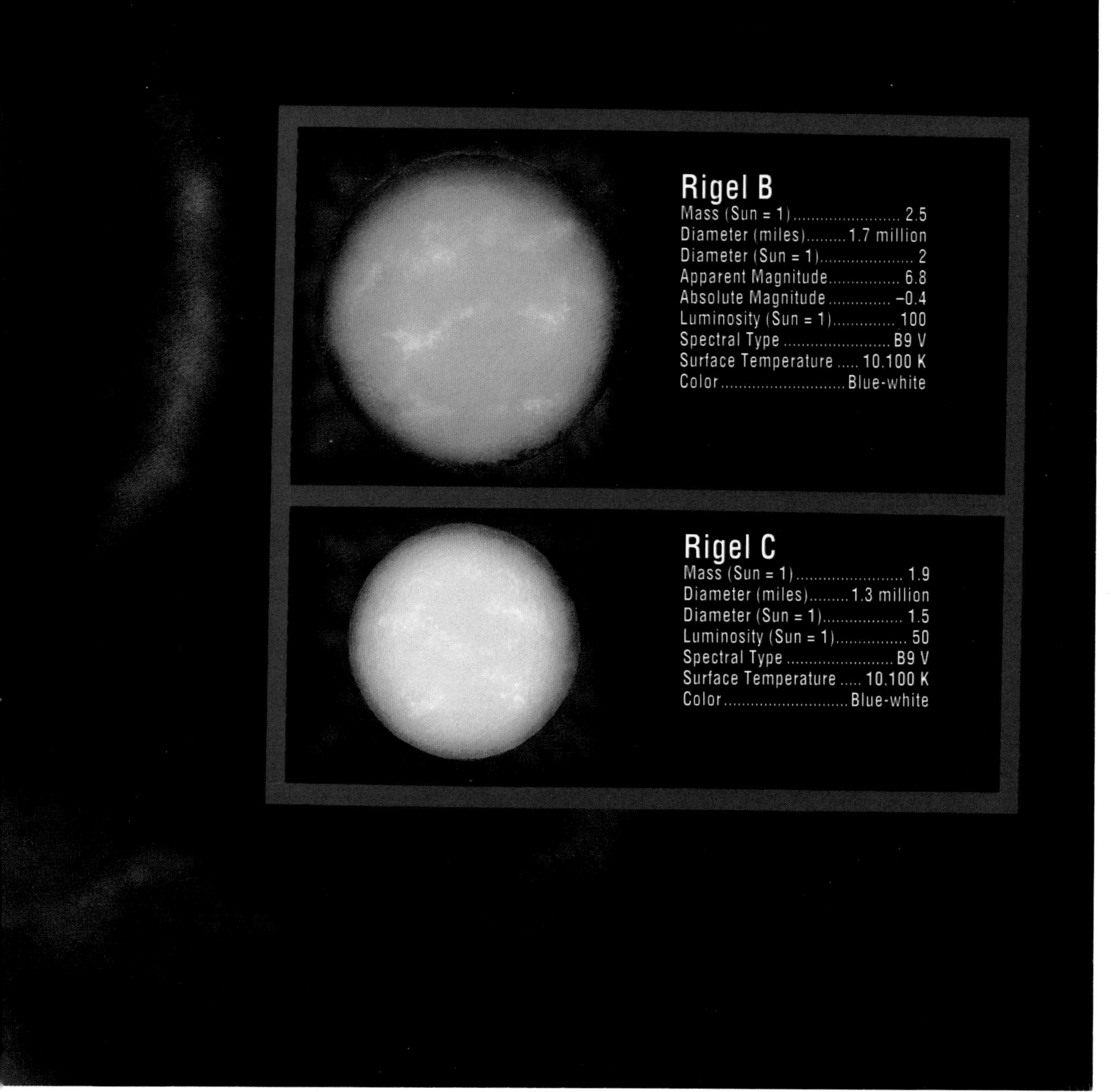

Energy radiates through the Sun's interior, but is transported to the surface by convection.

differently. The story picks up in 1937, when spectra of Rigel B were taken at Mount Wilson Observatory. These new observations showed much greater detail than any previous observations and clearly revealed two sets of spectral lines. To everyone's surprise, instead of Rigel A, Rigel B turned out to be a spectroscopic binary! Both stars are blue main-sequence stars and they orbit one another once every ten days. There was no question about these results, as there had been with Plaskett's conclusion regarding the primary star, because both sets of lines had been captured by the spectroscope.

Even that wasn't the end of the mystery of Rigel's multiplicity, however. Over the years many observers, including such famous double-star observers as S. W. Burnham and R. G. Aitken, reported seeing Rigel B as a double star on certain occasions. The observed separation of the two stars was very small — less than 0.2 arcsecond. It doesn't appear that these observations could be of the spectroscopic companion, however. The orbital elements calculated in 1947 for the spectroscopic pair showed that these stars never get even as much as 0.1 arcsecond apart. Other observations made with large telescopes on nights with excellent observing conditions also indicate that Rigel B is not a visual double. Like Plaskett, the observers of the visual companion to Rigel B seem to have been mistaken. As Roscoe Sanford, the astronomer who calculated the orbit of Rigel B's spectroscopic companion, diplomatically put it in 1942, "the double star and spectroscopic observations of Rigel's companion are still to be reconciled."

Enveloped in Gas and Dust

In addition to its immediate family, Rigel's interstellar neighborhood also contains a cloud of dust, known as IC 2118. Immense clouds like this one are the progenitors of new stars, which form as the clouds contract and fragment under the influence of gravity. Most of the Galaxy's dust and gas resides in the spiral arms of the Milky Way. Because so many young stars congregate in this area, we see a vast number of the massive, fast-burning blue stars when we gaze toward Orion.

Rigel is so luminous that it illuminates the dust in IC 2118 even though the nebula lies 2.5° away in the constellation Eridanus. Observers on Earth see the cloud as a dim reflection nebula. Edwin Hubble calculated the cloud to be some 20 light-years across. He wrote that "this seems an enormous distance over which to assume stellar radiation to be effective in illuminating nebulosity." His calculations indicated, however, that Rigel's situation was not unlike that of the nebulosity surrounding the Pleiades: We see blue starlight scattered in preference to red. This is the same process that makes Earth's daytime sky appear blue.

Just as there are many Pleiads, Rigel is not alone in supervising its stellar backyard. In the mid-1960s, Canadian astronomers Sidney van den Bergh and Rene Racine studied many such cases of stars in and near reflection nebulae. They grouped Rigel and seven other stars into what is known as a "T association," named after the star T Tauri, which belongs to the first such grouping that was identified. Rigel's association is called Taurus-Orion R1 and encompasses a far greater volume of space than even the cloud Hubble observed. It's easy to imagine that hundreds, and perhaps thousands, of additional stars belong to this association, but that they are too faint to observe from Earth.

Our view of Rigel has thus been expanded to include three stars and a colossal sphere of gas: a close double, Rigel B and Rigel C, that would just about fit on opposite sides of our solar system; the bright blue supergiant itself, a hundred times farther away than the distance between B and C, or about 23 light-days; and a reflection nebula more than 20 light-years across, with the whole Rigel system suspended just beyond the tenuous outer reaches of this vast bubble.

But even this is not the whole picture. Rigel A also has a considerable shell of gas around it, which has been the subject of study for several decades. In 1986 Daniel Hayes concluded that it is likely "Rigel undergoes periodic disturbances marked by an outflow (and possible inflow) of material." In other words, the stellar material in the outermost layers of Rigel is being "puffed" outward now and then. Some of this matter later falls inward again, back toward the surface of the star.

Hayes based his conclusion on ten nights of observing the polarization of the starlight from Rigel A. Polarization is the vibration of a photon at right angles to its direction of travel. As the photon moves forward at the speed of light, it can vibrate up and down, left

On Rigel, on the other hand, energy is transferred by convection in the interior and by radiation near the surface.

and right, or at any other angle. Ordinarily, a beam of light consists of a mix of all angles. If one particular angle or range of angles predominates, astronomers say the light is polarized. Because Rigel is so bright, its light can be scrutinized in great detail, and the degree to which its light is polarized can be accurately determined.

Hayes' observations confirmed that Rigel's starlight was indeed polarized. In itself, this is hardly unusual. Many stars, particularly those with strong magnetic fields at their surfaces, radiate polarized light. What was more surprising was that the polarization of Rigel's light varied from night to night. A detailed analysis of the observations failed to turn up any pattern to this variability. Furthermore, Hayes used the polarization data to calculate that strong magnetic fields are probably not responsible for the "puffs" in Rigel's circumstellar envelope. Thus Rigel A is different from most of the other stars whose light is polarized. Instead, Hayes suggested that pulsations on Rigel's surface might be the cause of the "puffs."

Most stars seem to lose mass via "stellar winds," but these "puffs" from Rigel A are rather peculiar. Estimates vary as to how much mass is lost this way, but it seems reasonable that Rigel A could lose about 1 percent of its mass in a million years. The star puffs off a shell of gas about 100 million kilometers above its outer atmosphere. Expanding at almost 200 kilometers per second, the shell of gas would reach Jupiter's orbit in a month if Rigel A were located at the center of our solar system.

How are these puffs ejected? Are they connected to asymmetric pulsations of the star's surface? Why do they occur irregularly, and what implications does this have for our understanding of the inner workings of stars in general? Why does some of the material fall back to the surface, unable to escape Rigel's gravity? Astronomers still don't have satisfactory answers to such questions.

An Explosive Future?

For astronomers, the topic of mass loss from blue supergiants similar to Rigel took on a new urgency when Supernova 1987A burst forth in the Large Magellanic Cloud four years ago. Astronomers soon determined that the supernova's progenitor star, Sanduleak –69°202, had been just such a star. The conventional wisdom had always been that red supergiants, like Betelgeuse, were the most likely candidates to become supernovae.

Challenged to explain how a blue supergiant had met this catastrophic fate, theorists adjusted their computer models. Much to their surprise, the models predicted that a blue supergiant in an advanced evolutionary stage could indeed become a supernova — once the surface had ejected a great deal of mass. Although the shell of gas surrounding Rigel A probably does not yet contain enough mass to make the star an immediate supernova candidate, this could well be the kind of behavior that Sanduleak –69°202 exhibited in the recent astronomical past.

Of all the millions of stars like Rigel, sprinkled around the spiral arms of the Milky Way and each one bright enough to be visible across spectacularly large distances, we can see but a few. The rest are lost to us behind the dust lanes in the plane of the Galaxy. So for a small fraction of the Galaxy's diameter, the blue supergiants each reign over their own dominion, yet they are but princes on a cosmic continent.

There's more here than just the raw material for a fictional galactic adventure story. The next time you gaze at Orion, imagine yourself looking over the shoulders of men like Struve, Aitken, and Plaskett, the early pioneers who tried to plumb Rigel's secrets with varying degrees of success. Peering with you are the current generation of researchers, like Hayes, who are seeking to unravel thin threads of evidence about the detailed physical structure of Rigel, its atmosphere, and its circumstellar shell. And, off to one side, are the astronomers of the apocalypse, theorizing about supernovae and the cataclysmic things that go bump in the night. Think of that the next time you spy Rigel and its cousins down the two-thousand-light-year-long tunnel of the Orion Arm, blazing away in all their short-lived celestial glory. □

Peter Jedicke is an enthusiastic amateur astronomer who teaches math and physics at Fanshaw College in London, Ontario. David H. Levy is an author and comet hunter, with six comet discoveries currently to his credit.

A Star That Breaks All the Rules

A nearby, fast-spinning red dwarf star named Gliese 890 has astronomers puzzling over its mysterious origin and erratic behavior.

by Ken Croswell

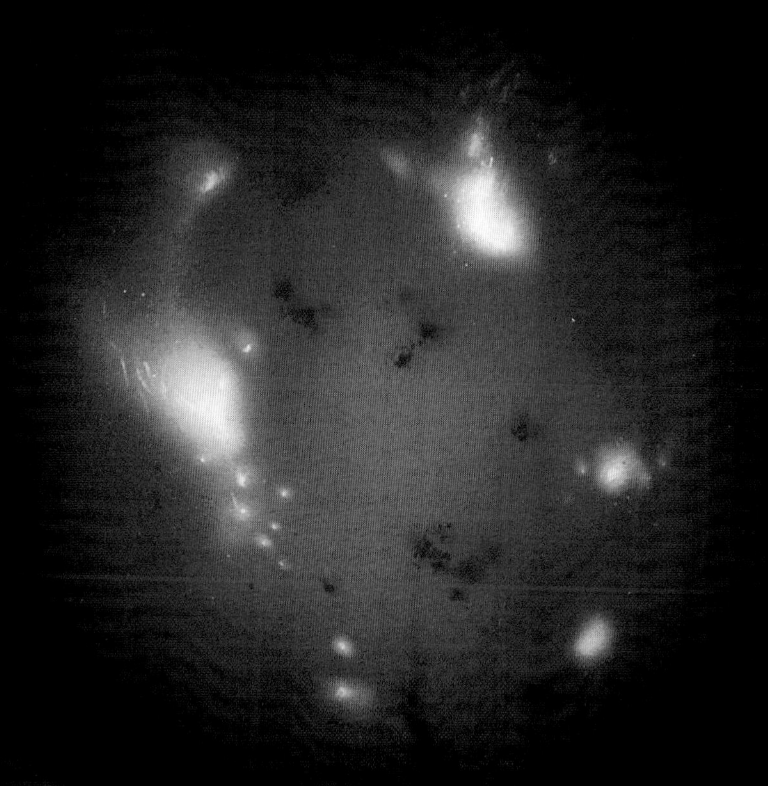

Faint, cool, and small, most red dwarf stars lead quiet lives of little renown. Recently, however, a nearby red dwarf named Gliese 890 has intrigued and puzzled astronomers. They have discovered the star is spinning extremely fast — dozens of times faster than the Sun.

So what's the big fuss about a fast-rotating star? According to theory Gliese 890 should rotate very slowly, like other stars of its type — but it doesn't. And the more astronomers have looked at this star, the more oddities they have uncovered. Tracking down the behavior and origin of Gliese 890 has progressed from a few observations of an ordinary star to a full-blown astronomical mystery. How can a star with such an ordinary appearance and place in the Galaxy violate the rules that govern how stars behave? Astronomers set out to discover the answer.

Like the Sun, and unlike Gliese 890, most single late-type stars rotate slowly. They are cool yellow, orange and red stars of spectral types G, K, and M. (In contrast, many early-type stars — hot blue and white stars with spectral types of O, B, and A — spin fast.) A late-type star spins slowly because it has a magnetic field that carries gas away from the star's surface. As it loses more and more material, the star turns more and more slowly, just as a spinning ice skater spins more slowly when he extends his arms. Astronomers call this spin-down process magnetic braking, and it is responsible for the slow rotation of single, late-type stars.

Such is not true, however, for late-type stars in binary systems. In this case late-type stars can continue to spin fast, because they convert the spin energy of their orbits into rotational energy, thus counteracting magnetic braking. Each star in such a binary raises tides on the other, and the two stars get locked together.

But the binary star case cannot explain Gliese 890. This star is solitary and it rotates in less than half a day — a far cry from the Sun's rotation period of a month. The star's rapid rotation produces some curious by-products that fascinate astronomers who observe Gliese 890. These include starspots and stellar flares that cause the star's brightness to vary. The erratic behavior of this star has also started astronomers thinking about the star's origin.

"Gliese 890 is a rapid rotation laboratory for late-type stars," says Arthur Young of San Diego State University, who has studied Gliese 890's starspots. "But why is it rotating so rapidly? When these things are in binaries, that's not an issue — they have no choice but to rotate rapidly. But when they're single like this, then our thinking is that this is a signature of youth. But if the star is very young, then what's it doing in the solar neighborhood?"

Out of Obscurity

The recent interest in Gliese 890 marks a sharp departure from years of neglect. American astronomer Alexander Vyssotsky announced the star's discovery in 1952, but he buried it in a list of over a hundred other K and M dwarfs. Gliese 890 glows faintly at magnitude 11 and lies in eastern Aquarius. In 1957 German astronomer Wilhelm Gliese published his first *Catalogue of Nearby Stars,* which contained 915 stars within 65 light-years of the Sun. At a distance of approximately 65 light-years, star number 890 barely made it into Gliese's catalog.

There, for two decades, Gliese 890 sat. It was just another red dwarf in a Galaxy full of them. Red dwarfs are fainter and cooler than the Sun because they have less mass, but they outnumber all other types of stars put together: 70 percent of the stars in the Galaxy are red dwarfs.

In the late 1970s Arthur Young and his colleagues took spectra of 72 nearby red dwarfs, most drawn from Gliese's catalogue. Spectra can determine how fast a star turns, because rotation broadens the lines in spectra. One limb of a spinning star recedes from us and is redshifted, while the other limb approaches us and is blueshifted. The net result is that rotation causes every spectral line to appear smeared out. The faster the rotation, the broader the lines. By measuring the breadth of the lines, an astronomer can determine a star's rotation speed.

Gliese 890 has broad spectral lines and so is spinning extremely fast for a single, late-type star. The Sun rotates at just two kilometers per second, and most single red dwarfs in the solar neighborhood spin at less than five kilometers per second. Moreover, all single red dwarfs in the solar neighborhood rotate at less than 25 kilometers per second — except for Gliese 890. Gliese 890 is spinning at the extraordinary speed of 70 kilometers per second.

"We could not believe that an M dwarf star could rotate that rapidly and be single," says Young, who for many years assumed the star must be part of a binary system.

But Gliese 890 proved to be single: its line-of-sight velocity did not

Did Gliese 890 Escape the Pleiades?

Astronomer Arthur Young believes that Gliese 890 was once part of the Pleiades star cluster and was ejected toward us in space.

Here's how the scenario works: millions of years ago Gliese 890 was a typical red dwarf in the Pleiades. The star was gravitationally bound to the cluster and slowly orbited the Galaxy within the group.

Carried Away by Gravity

Several million years ago, Young believes, a massive molecular cloud passed close by the Pleiades and dispersed many of its members throughout surrounding parts of the Galaxy. This explains the Pleiades' anomalously low mass. Carried by the gravitational tug of the cloud, Gliese 890 headed toward us in space.

A Lone, Fast-Spinning Star

Still fleeing from its original birthplace, Gliese 890 today is a distant relative of the stars that still inhabit the Pleiades. The legacy of Gliese 890's past can now be measured in the star's fast rotation period, active magnetic field, occasional huge flares, and large starspots.

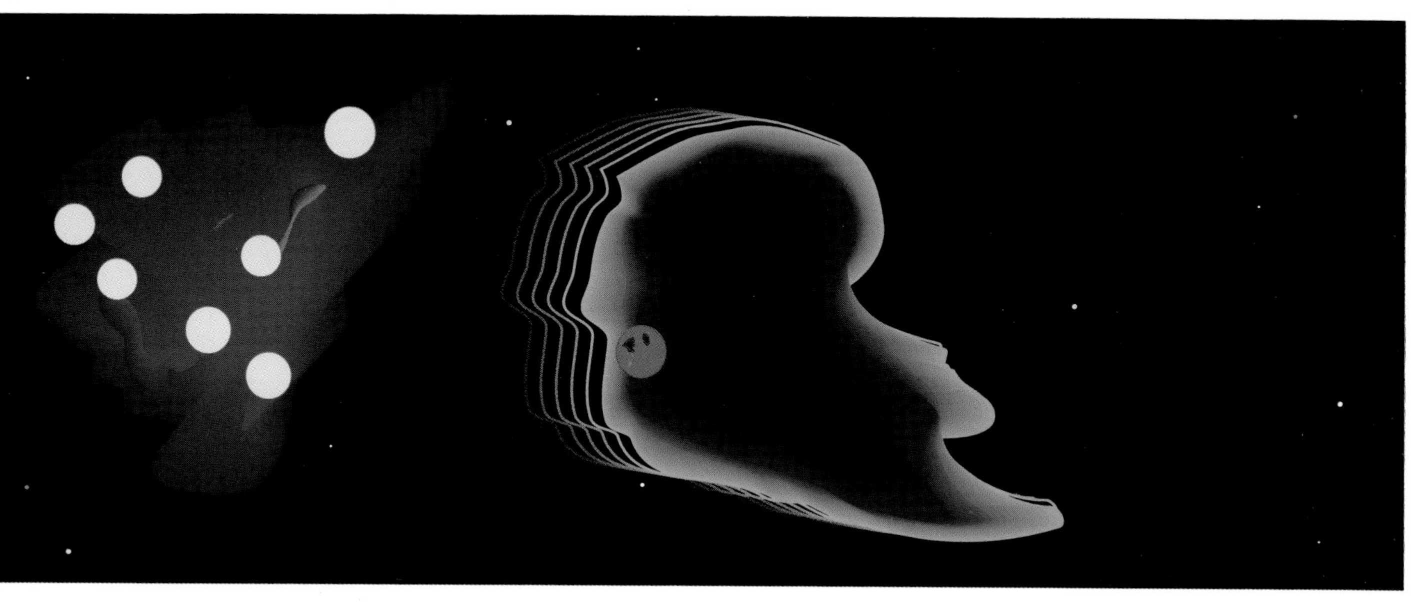

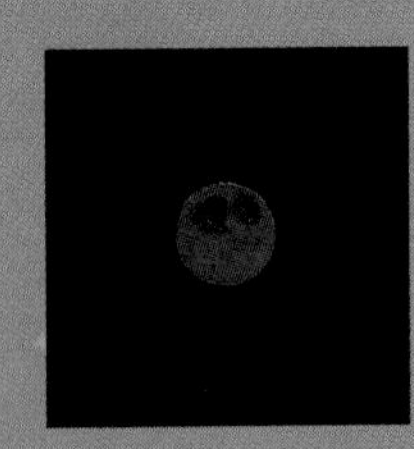

Lightning-Fast Rotation

Astronomers have determined that Gliese 890 is rotating at the astonishingly fast rate of 70 kilometers per second, 35 times faster than the Sun. The discovery was buried in the star's spectrum. The rapid rotation causes lines in the spectrum to broaden relative to other stars of the same type. This spectrum of absorption lines in neutral calcium shows that relative to the spectrum of Gliese 880 (top), different types of spectra of Gliese 890 are all much broader. Diagrams by Patti L. Keipe after Young, *et al.*

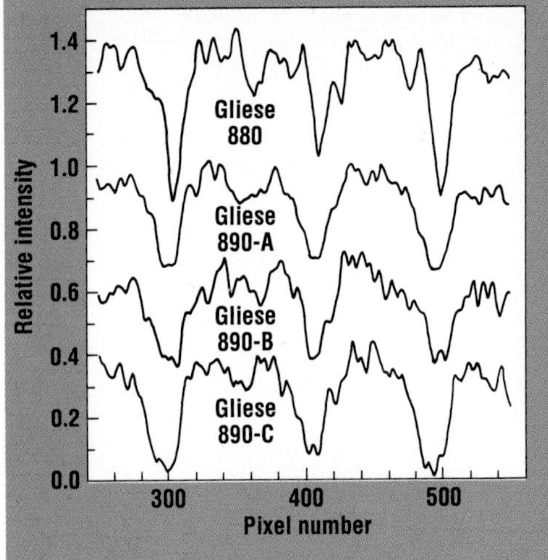

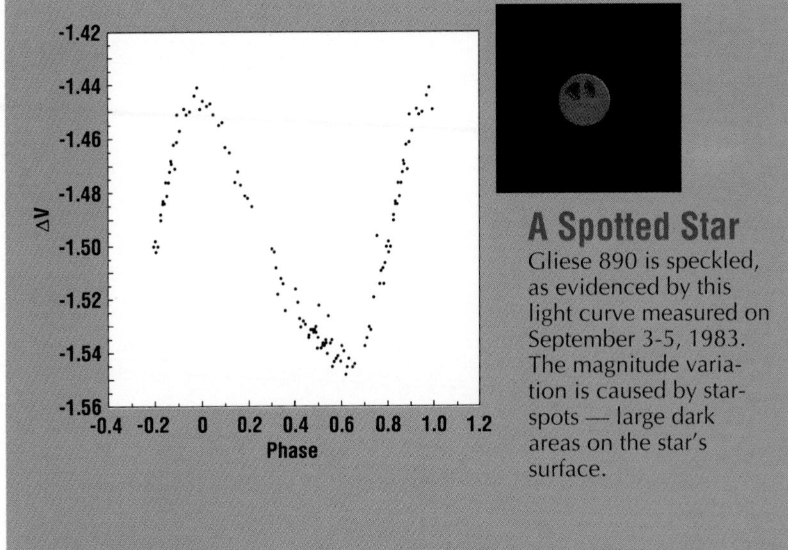

A Spotted Star

Gliese 890 is speckled, as evidenced by this light curve measured on September 3-5, 1983. The magnitude variation is caused by starspots — large dark areas on the star's surface.

A Star with a Temper

Curiously, for such a cool star, Gliese 890 has a hot temper. Erratically and without warning, the star has flared in brightness by a factor of two over time spans of just a few minutes. This diagram shows the brightness of Gliese 890 during a flare that was observed in 1983.

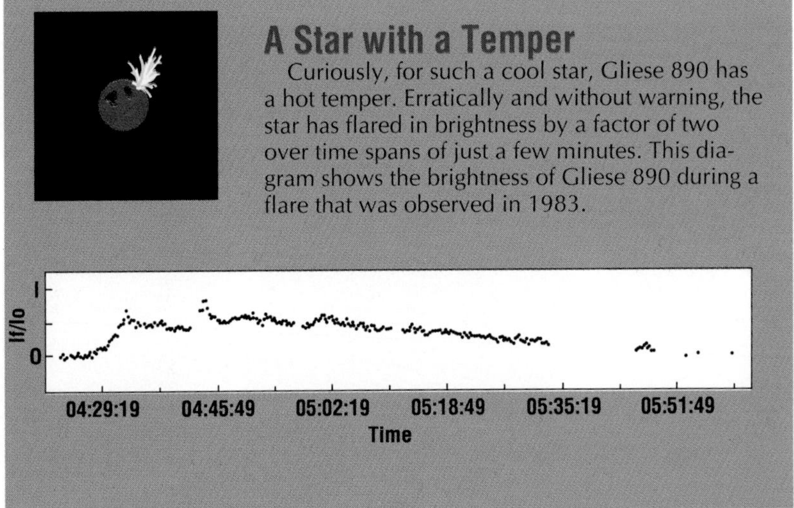

vary periodically the way it would if the star were revolving around a companion. Gliese 890 was spinning fast by itself; it had not been spun up to its high rotational velocity by a companion.

Astronomers began to wonder if there might be another way to approach the problem. In 1983 Scott Temple, now of the University of Washington in Seattle, found that the star's brightness varied by about a tenth of a magnitude every 10.3 hours. This variability is caused by starspots that come in and out of view as the star spins and means that the star rotates in just 10.3 hours.

"That was a wonderful confirmation," says Young, who had calculated what the rotation period should be if the star were spinning at 70 kilometers per second. "We predicted something like eight or nine hours, and the actual number my student, Scott Temple, found was ten hours."

Starspots and Stellar Flares

The starspots that cause Gliese 890's light to vary arise from the star's strong magnetic field, itself a product of Gliese 890's rapid rotation. Astronomers believe that a star needs convection and rotation to have a strong magnetic field. Gliese 890 has plenty of both.

Convection means that the star transports energy by bubbling and boiling. Hot gas rises to the top of the star, gives up its heat, then sinks back down again to collect more heat. The outer layers of all stars redder than spectral type F5 are convective. Astronomers observe such convection on the Sun's surface, and it presumably occurs on Gliese 890 as well. When the gas in these late-type stars rises, it carries magnetic field lines with it.

But convection alone is not enough to produce a strong magnetic field. The other requirement is rotation, which governs the convection and strengthens the field further. Stars like the Sun rotate differentially: the Sun's equator spins faster than its poles. A magnetic field line that stretches initially straight from the Sun's pole to its equator gets wound around and around as the Sun rotates, intensifying the magnetic field. The faster the rotation and greater the differential, the stronger the magnetic field gets. In a slowly spinning star like the Sun, with its rotation speed of two kilometers per second, the magnetic field is weak and the activity it produces is mild. Sunspots occur over a small fraction of the solar surface and produce occasional solar flares. A fast-spinning star — one with a rotational velocity of five or ten kilometers per second — has a stronger magnetic field and so many starspots that the light of the star fluctuates as the star turns.

Rotating at 70 kilometers per sec-

Gliese 890 and the Sun

	Gliese 890	The Sun
Right Ascension (2000.0)	23h08.3m	—
Declination (2000.0)	–15°25'	—
Distance	65 ly	—
Apparent Magnitude	10.9 (v)	–26.72
Absolute Magnitude	9.3	4.9
Relative Brightness	0.017	1.0
Spectral Type	M2	G2
Color	Red	Yellow
Temperature	3600 K	5770 K
Relative Mass	0.5	1.0
Relative Radius	0.6	1.0
Rotation Speed	70 km/sec	2 km/sec
Rotation Period	10.3 hrs	One month

ond, Gliese 890 must have a very strong magnetic field. Gliese 890's starspots are one consequence of this strong magnetic field. K and M dwarfs that vary in brightness because of starspots are called BY Draconis stars, after their prototype star in the constellation Draco. Many BY Draconis stars rotate faster than five kilometers per second and exist in binaries whose companions have spun them up. The rapid rotation produces the magnetic fields that cause the starspots that cause the variability. As in the case of Gliese 890, astronomers can measure the rotation periods of these stars by determining the periods of the stars' variability.

Another consequence of Gliese 890's rotation-driven magnetic field is huge stellar flares. In 1983 Bjørn Pettersen and colleagues of the University of Oslo discovered a gigantic flare on Gliese 890 that doubled the star's ultraviolet brightness in just a few minutes. Many other red dwarfs emit flares, too, and astronomers call these objects flare stars.

Did Gliese 890 Escape from the Pleiades?

Even as they study how rapid rotation affects Gliese 890, astronomers wonder how such an extreme star arose in the first place. The star must be very young, since it has not had time to brake its rotation. Yet, unlike other young stars, it's out there all by itself, rather than part of a star formation region like the Orion complex or a young cluster like the Pleiades.

Star clusters such as the Pleiades have told astronomers a lot about how stellar rotation decreases with age and provide strong evidence that Gliese 890 is very young. John Stauffer of the Harvard-Smithsonian Center for Astrophysics and others have found many fast-spinning, late-type stars in the Pleiades cluster. In fact, some K and M dwarfs in the Pleiades spin as fast as Gliese 890. The Pleiades is only 70 million years old, so its stars have not yet had time to spin themselves down.

Perhaps Gliese 890 is telling us that our theories of star formation are wrong. Astronomers believe that young stars form with thousands of other young stars in huge clouds of gas and dust. When the gas and dust disperse, the stars remain together in a star cluster like the Pleiades, which itself eventually disperses.

But perhaps all stars do not form in this way. Perhaps some, like Gliese 890, are born alone, without any neighbors. Gliese 890 may have started its life when a small, isolated cloud of gas and dust began to contract many millions of years ago. This small gas cloud formed a single, small, rapidly rotating star. The gas and dust dispersed, leaving us with a young isolated star that now casts doubt on our ideas concerning star formation.

Young prefers a different scenario, however. He notes that Gliese 890 is just like the rapidly rotating K and M dwarfs in the Pleiades. Although the Pleiades are farther from us than Gliese 890 — the cluster's distance is 410 light-years versus Gliese 890's distance of 65 light-years — Young thinks that Gliese 890 was actually once part of the Pleiades itself. Millions of years ago, though, the star somehow got ejected in our direction. "The space velocity of Gliese 890 fits the Pleiades remarkably, with only a small difference, which would be just what you'd expect if it were kicked out of the Pleiades," says Young. "Gliese 890 is nowhere near the Pleiades in the sky, of course — it's in the southern hemisphere. But that doesn't matter, because the Pleiades aren't that far away, and with a few kilometers per second velocity, Gliese 890 can make it to our part of the Galaxy in a few million years."

Young speculates that the Pleiades was once a much greater star cluster. He believes that a giant molecular cloud weighing a million times more than the Sun may have passed close to the Pleiades. The gravity of this huge cloud tore loosely bound stars from the cluster, including Gliese 890.

"One thing that is very striking about the Pleiades is that it's an anomalously small cluster," he says. "Open clusters typically have about a thousand times the mass of the Sun. But the Pleiades have perhaps two or three hundred times the mass of the Sun." This suggests that the original Pleiades star cluster may have lost over half its stars.

"I would guess that there are stars all over the sky that are former members of the Pleiades," says Young, who points to the star AB Doradus as another example. AB Doradus is a nearby, K-type star that rotates in just half a day. The star's strong magnetic field produces starspots and stellar flares, just as on Gliese 890.

"I think both of them are escapees from the Pleiades," he says. "There's a growing consensus that a lot of these low-mass, rapidly rotating stars we see are likely to be connected to the Pleiades — not all of them, but a lot of them."

Whatever its origin, Gliese 890 offers astronomers the rare chance to study rapid rotation in a late-type single star, rotation that produces magnetic activity such as starspots and stellar flares. But the star's peculiar nature also suggests that Gliese 890 has had a most unusual history — a history that may involve one of the most beautiful star clusters in the sky. □

Ken Croswell received his doctorate in astronomy from Harvard University for studying the Milky Way Galaxy. He has written for Astronomy Now, New Scientist, *and Time-Life Books.*

Will Supernova 1987A Shine Again?

As strange as it may seem, this recent supernova may provide more fireworks in the years ahead.

by Laurence A. Marschall and Kenneth Brecher

SUPERNOVA 1987A BRIGHTENS AGAIN as the debris from the 1987 explosion collides with a pre-existing ring of gas, causing it to glow. Illustration by Robert Eggleton.

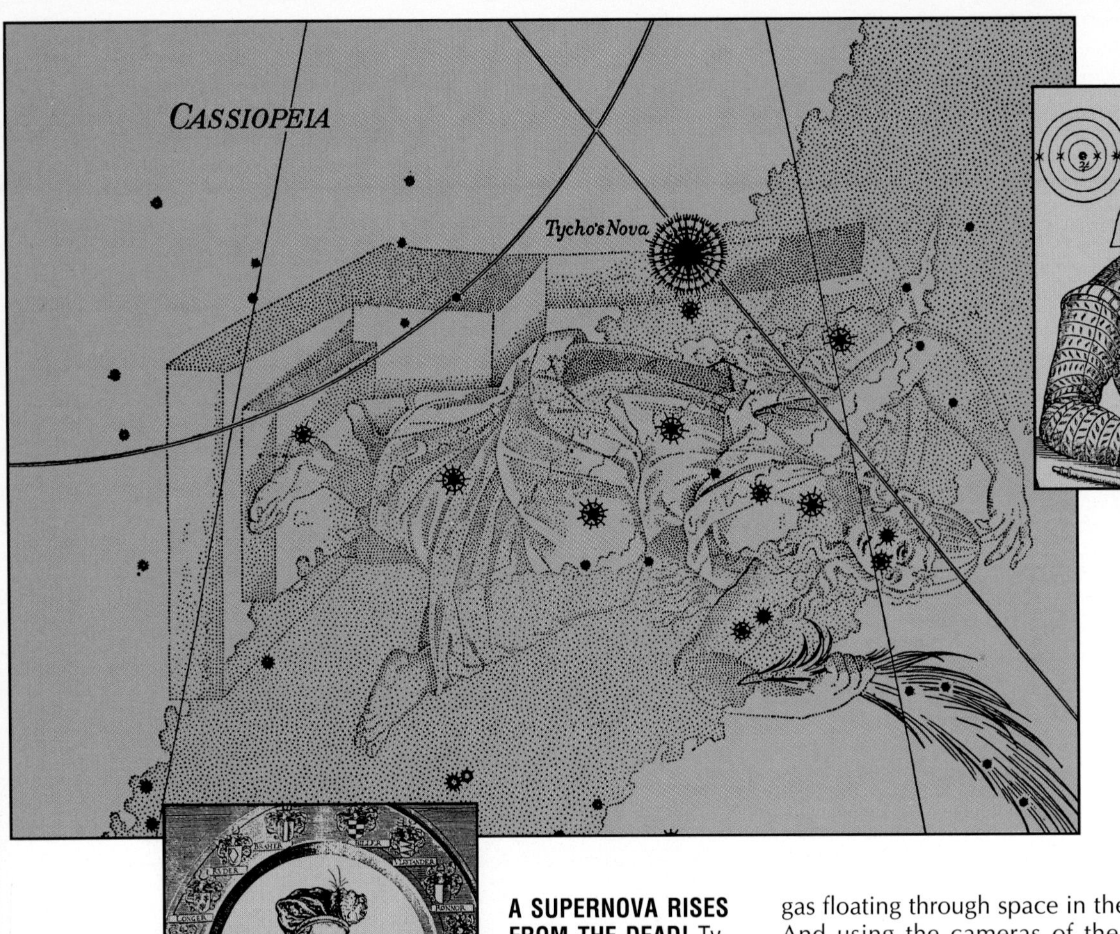

A SUPERNOVA RISES FROM THE DEAD! Tycho (left) observed a bright supernova in 1572 in Cassiopeia (as depicted in Johann Bayer's *Uranometria*, above). Forty years later, in 1612, Simon Marius (top right) observed the supernova's apparent rebrightening.

The reports of my death have been greatly exaggerated. — Mark Twain, in a cable to the Associated Press, 1897.

Have reports of the death of Sanduleak –69°202 been exaggerated? This blue supergiant star burst onto the astronomical scene five years ago, on February 24, 1987, as a brilliant point of light visible in the southern sky. Sanduleak –69°202, until then an obscure member of our neighboring galaxy, the Large Magellanic Cloud, had come to a sudden and violent end. The resulting explosion, called Supernova 1987A, has kept astronomers jumping ever since. Because it was the brightest supernova in almost four centuries, SN 1987A provided an unprecedented view of the last days of a massive star, from the exhaustion of its supply of nuclear fuel, through the catastrophic collapse of its innermost core, to the ultimate shedding of its outer layers in a cloud of outrushing debris.

In the months and years following Supernova 1987A, astronomers have been treated to a host of remarkable events. They detected neutrinos emitted from deep inside the dying star, where temperatures reached into the billions of degrees. (See "Insight into Star Death" by Richard Talcott, February 1988.) They photographed faint echoes of light that had ricocheted off clouds of dust and gas floating through space in the vicinity of the blast. And using the cameras of the Hubble Space Telescope, they discerned the shape of the ragged debris from the explosion as it expanded outward.

Now the fireworks are over. What is left of SN 1987A shines far more dimly than Sanduleak –69°202 ever did. In fact, if we did not know its history, the supernova would be an inconspicuous dot among the hundreds of millions of stars in the Large Magellanic Cloud.

Old supernovae, however, may not just fade away. Like the final, delayed pop of a Fourth-of-July rocket, the final fireworks from a supernova may come years after the initial explosion. Given what we've learned from the behavior of supernovae in the past, it's quite possible that SN 1987A will be rejuvenated, shining again for a brief period at roughly 9th magnitude sometime within the next decade.

Shine again? The very thought seems strange. In the act of becoming a supernova a star is torn to shreds, leaving behind only a small, dense core of neutrons (and often, not even that). Because the body has been cremated and the ashes scattered, nothing remains to rise from the dead. Though variable stars of other sorts may brighten and dim more than once, supernovae, one would think, should have but one brief moment of glory before disappearing forever.

Clues from the Past

That's why it is surprising to discover reports of supernovae coming back to life long after they had faded from view. A remarkable account uncovered by one of us (KB) appears in the writings of Simon Marius, astronomer to the court of Brandenberg at the beginning of the 17th century. Marius was a model Renaissance scientist — colorful, clever, and

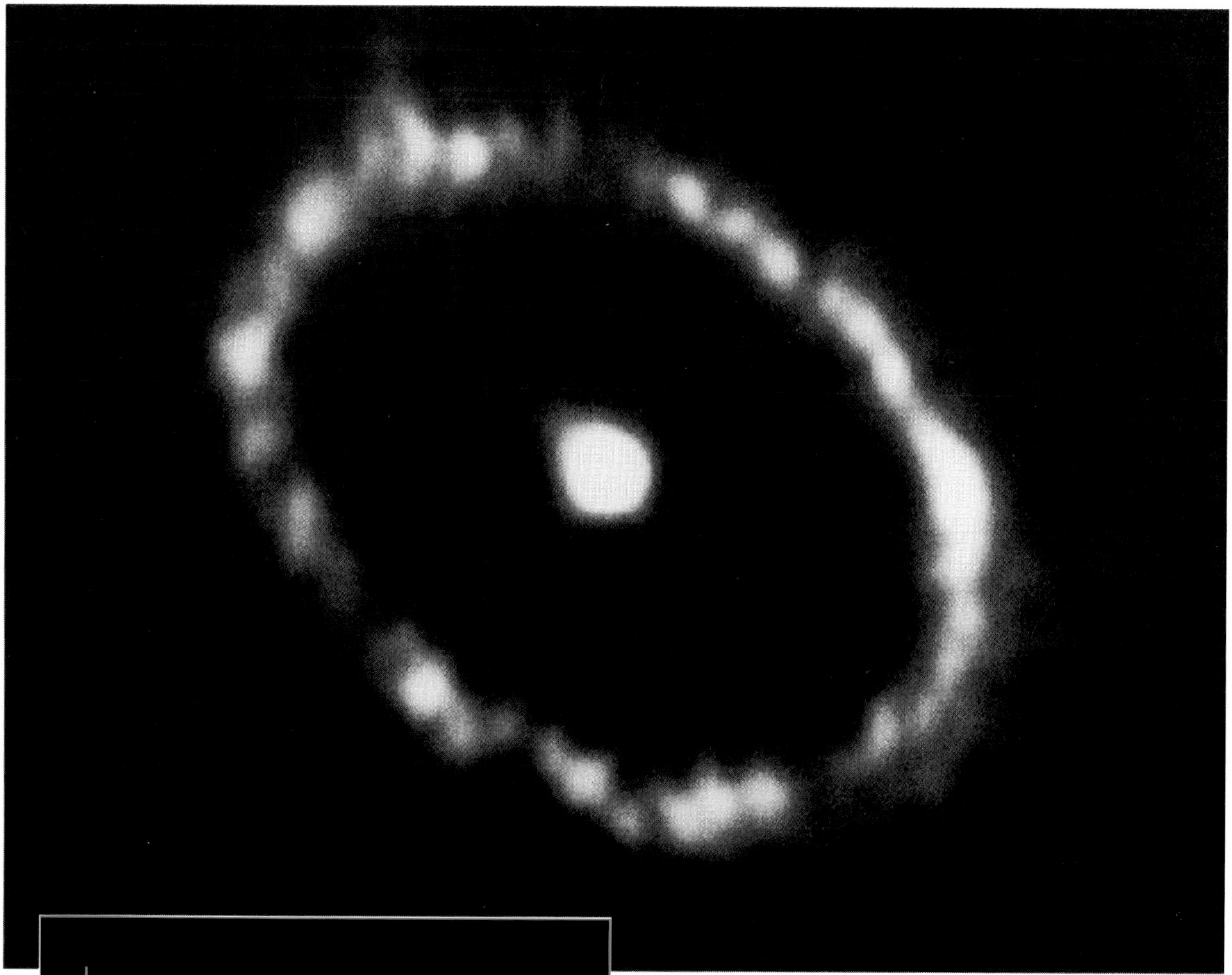

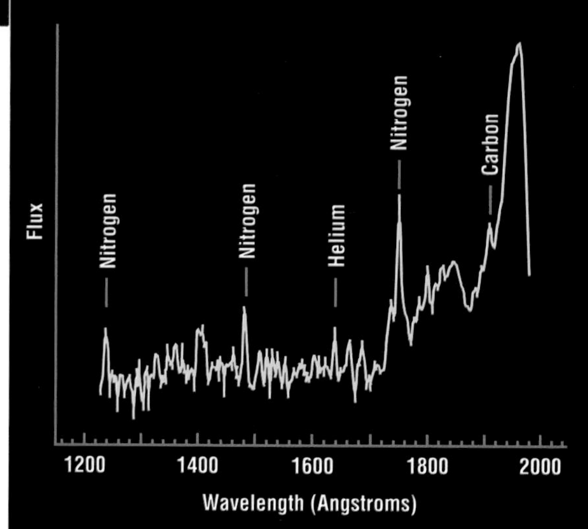

THE FIRST INKLING that a ring surrounded Supernova 1987A came from an IUE spectrum (left) that showed strong lines arising from nitrogen in the ring. The ring appeared in all its glory to the Space Telescope's Faint Object Camera (above). Spectrum courtesy Robert Kirshner; photo courtesy NASA/ESA.

quick to adopt new tools of the trade. Using the newly invented telescope (which Marius immodestly claimed to have invented himself) he observed Jupiter's four bright moons and published the first sightings of the Great Nebula in Andromeda.

Some of Marius' more extravagant claims angered his more famous contemporary, Galileo Galilei. Incensed by what he considered an encroachment on his territory (he had published the first observations of Jupiter's moons), Galileo called Marius a "poisonous reptile" and an "enemy . . . of all mankind." But modern scholars have praised Marius' meticulous recordkeeping, and his names for the moons of Jupiter — Io, Europa, Ganymede, and Callisto, which were taken from Greek mythology — have long outlasted Galileo's proposal that the moons be named "The Medician Planets" after his patron, Cosimo de Medici.

Marius had worked briefly with Tycho Brahe, the great Danish astronomer. He knew that in the year 1572 Tycho had discovered a bright new star, a "Nova Stella" in Tycho's writings, in the constellation Cassiopeia. In a short book he wrote about the new star, Tycho included a map showing the star's position near the "W" of Cassiopeia, along with a detailed account of the star's behavior. When it first appeared, Tycho noted, the star outshone the planet Venus, and it remained visible to the naked eye for a full two years.

From Tycho's detailed description, modern-day astronomers recognize that the "star" of 1572 was actually an exceptionally bright supernova. Using X-ray and radio telescopes, astronomers can detect a faint remnant ring of gas lingering around the site of the explosion. However, there is no trace of a star at its center.

During Tycho's era, of course, no one even

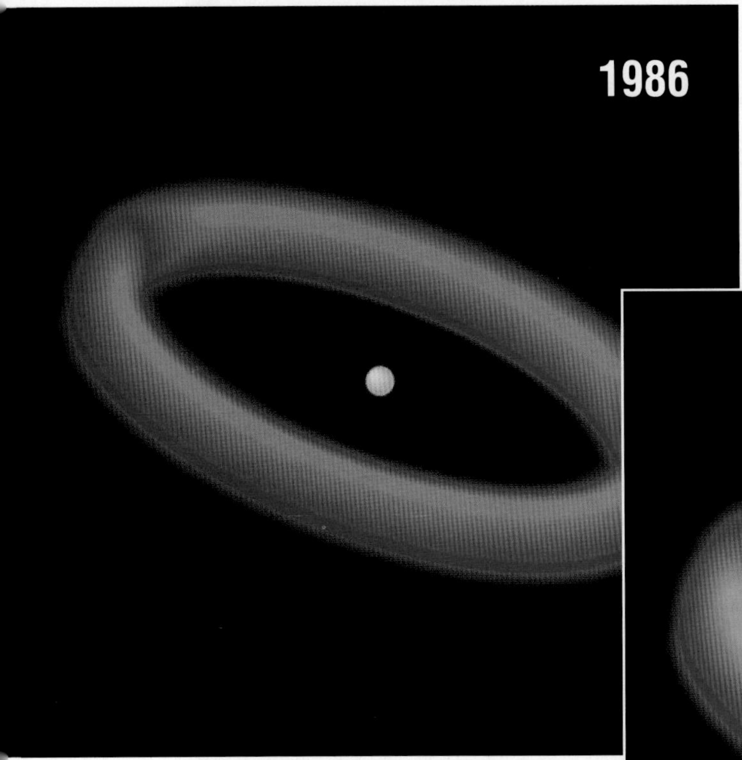

1986

How it will happen

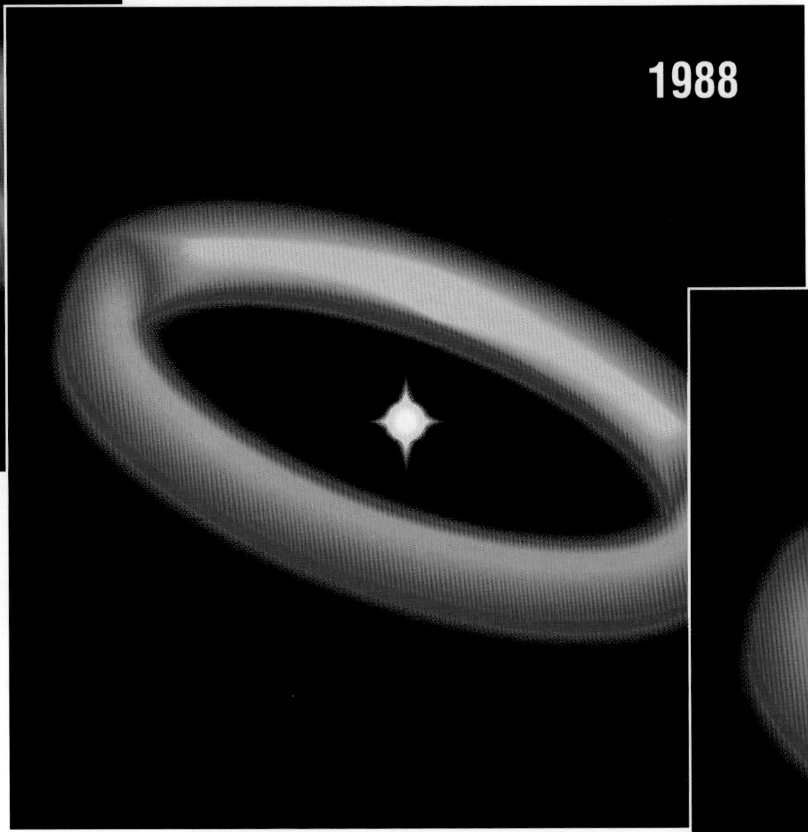

1988

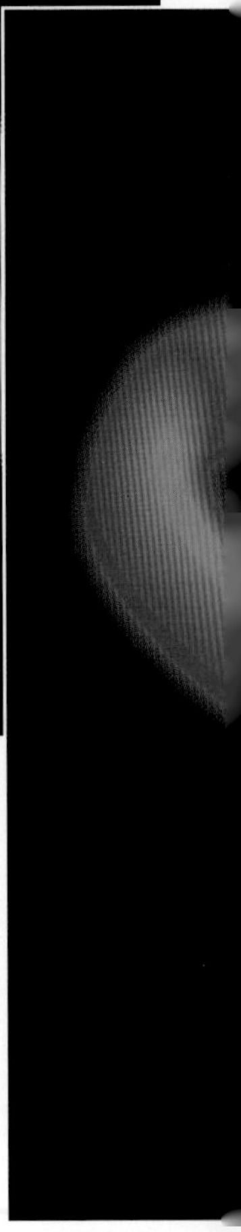

dreamed that a star could self-destruct, and no reason existed to expect that the nova of 1572 would disappear forever. Tycho, in fact, believed he was witnessing the birth of a star, not its death, and so it was natural for later astronomers to look for it with better instruments. Among those astronomers was Simon Marius. On clear nights in the winter of 1612-1613, wrote Marius in a book called the *Prognosticon*, he caught sight of the star that Tycho had last seen four decades before. It was nowhere near as brilliant as it had been, to be sure, and if the telescope had not been invented a few years earlier, it might not have been noticed at all.

Marius described the star as somewhat dimmer than Jupiter's third moon, Ganymede, and remarked that it disappeared from sight after a year's time. This would place it somewhere between 5th and 6th magnitude, making it almost 10 magnitudes — 10,000 times — dimmer than at Tycho's original sighting. But Marius was certain that the star's position coincided exactly with the mark on Tycho's map of 1572. Modern star maps show no star of comparable brightness within a degree of the 1572 supernova, so it seems unlikely that an observer of Marius' caliber would have mistaken a nearby star for Tycho's.

Evidence of other supernova reappearances can be found in the meticulous astronomical records kept for the emperors of China. According to scholar Wan Jian-Ming, the ancient chronicles tell of more than one such reappearance. The supernova of 1006, the most brilliant supernova ever recorded, shone forth again in the year 1016, almost 10 years to the day after its initial sighting. Chinese astronomers, even at the time, were struck by the coincidence, though there are no details as to how bright the star appeared at the second sighting. And there is another account that suggests the supernova of 1604, which was recorded by both the Chinese and the Europeans, reappeared in 1664. In both years, the Chinese records report a "guest star," or supernova, near the edge of the constellation they called Wei-Suei. (Today we know this region as Ophiuchus.)

Is There Life after Death?

What are we to make of these reports? Observers can be mistaken, of course. Even careful observers like Marius could have been influenced by the desire to see what they expected to see — in his case a new star just beginning to shine. And Chinese astronomers sometimes embellished eyewitness accounts with fanciful stories of sightings that provided astrological support for the adventures of their royal patrons. Thus the reappearances of supernovae in historical times might be no more substantial than sightings of Elvis reported in supermarket tabloids.

But current astrophysics suggests that supernovae have more of a chance of coming to life again than Elvis does. Shortly after the explosion of Supernova 1987A, these historical reports led one of us (KB) to predict a temporary revival of SN 1987A's light output sometime within the next decade.

The light will come, we believe, not from the star that exploded, but from the outrushing shell of debris the star ejected into the surrounding space. Immediately after the blast in 1987, most of the original mass of Sanduleak −69°202 was shot into space at speeds as high as one-tenth the speed of light. The material is still coasting outward today, an expanding shell containing about 15 times more mass than that in our own star, the Sun.

Were there nothing standing in the way, the debris from the supernova would thin out and disappear without a trace. Over 100,000 years or so, it would dissipate into the surrounding interstellar medium, becoming indistinguishable from the thin wisps of matter that drift between the stars. But such is not the case. The expanding shell of debris is on a collision course with another cloud of debris, material shed from Sanduleak −69°202 long before the explosive events of that star's death.

For probably a million years prior to the supernova explosion, Sanduleak −69°202 led the life of an aging star. As it consumed its internal stock of nuclear fuel, its diameter expanded a thousandfold and its surface cooled. It was a red supergiant. Later, in the last hundred thousand years before it died, the star shrank and heated up again, becoming a blue supergiant.

Both red and blue supergiant stars boil off large amounts of gas from their surfaces, a process astronomers call a "stellar wind." These winds blow far faster than the gales of planet Earth, however. Red supergiant winds flow into space at a few thousand miles an hour, while the winds from the hotter blue supergiants are about 100 times faster. Though neither process is anywhere near as violent as a super-

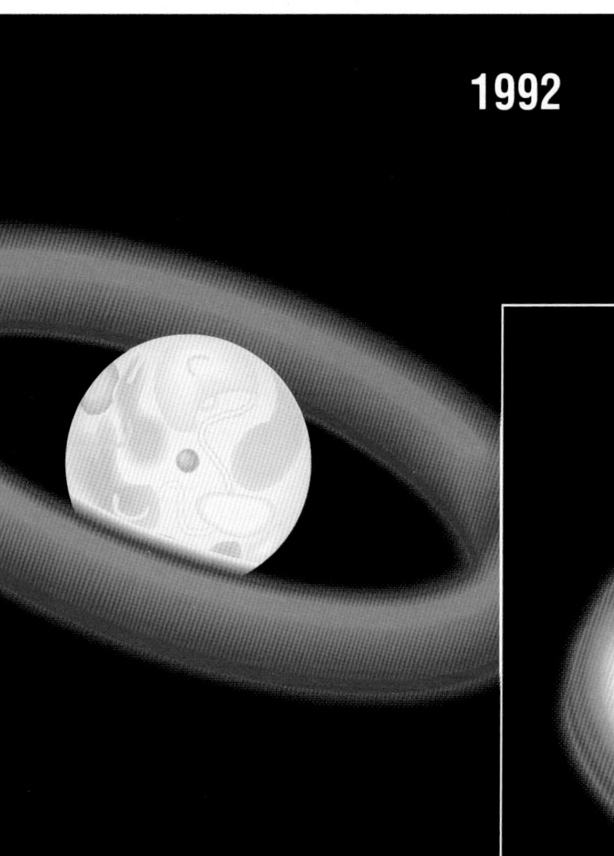

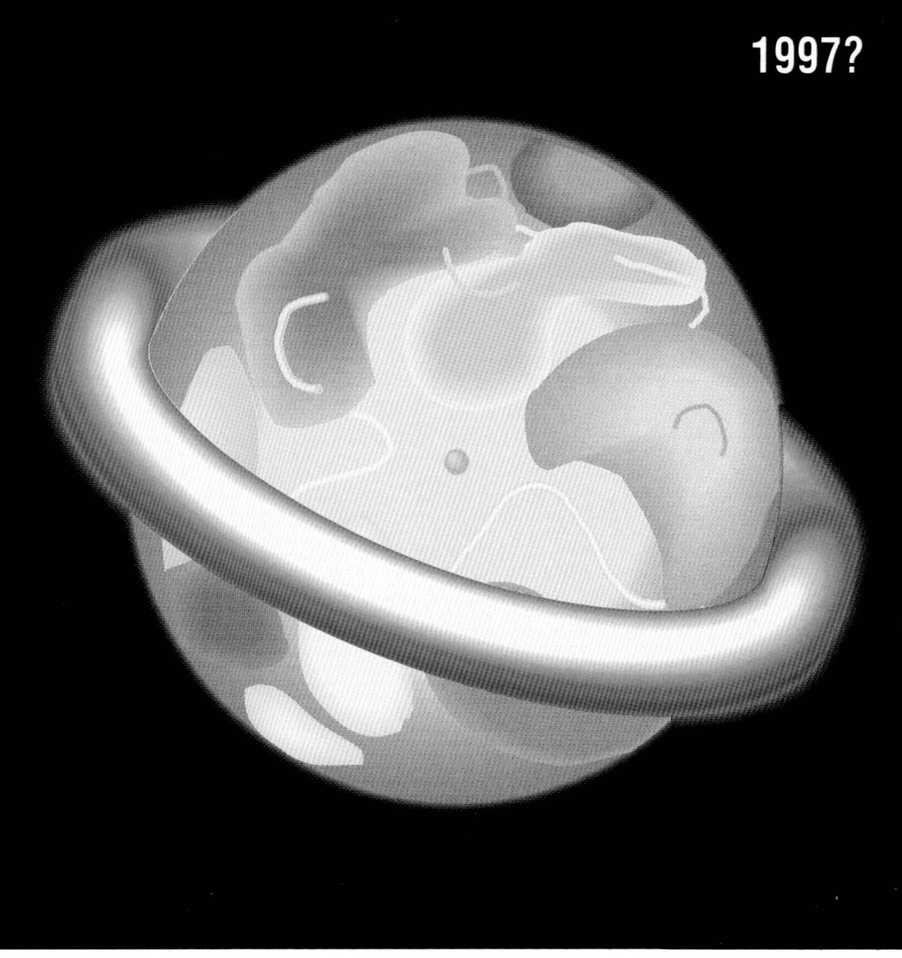

OUTRUSHING DEBRIS hits a surrounding ring of gas in this artist's interpretation of how the region of 1987A may look. In 1986 the progenitor star lay at the center of a gaseous ring. In 1988 the ring lit up slightly when 1987A's light reached it. Now the debris is about halfway to the ring. In a few years, the expanding shell will collide with the ring, and 1987A is predicted to shine again.

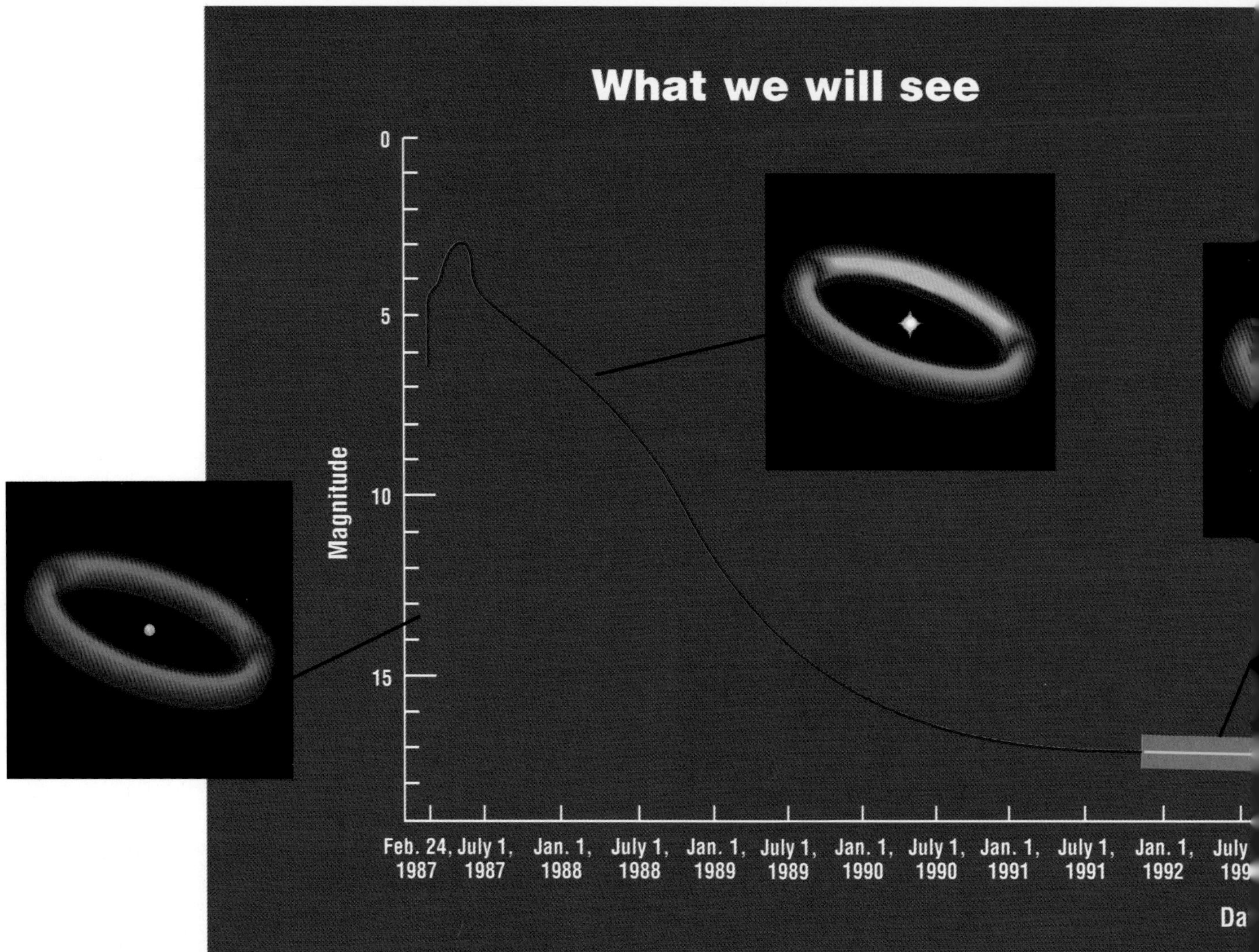

nova explosion, a substantial cloud of gas can be puffed into space during the lifetime of a supergiant star.

In the case of Sanduleak –69°202, the red supergiant wind carried off several solar masses of material in about a million years. Then in the star's last 100,000 years of life, its blue supergiant wind began blowing, piling up the material from the red supergiant wind that preceded it like a snowplow pushing into a snowdrift. By the time Sanduleak –69°202 was ready to explode, the plowed-up material from the stellar winds formed a ring of gas and dust about a light-year in diameter surrounding the site of the impending supernova.

This circumstellar ring now stands in the way of the expanding debris from SN 1987A. When the cloud of debris hits the ring, sometime within the next decade, the ring will heat up and begin to glow. To observers on Earth the supernova will shine again.

A Shell Game

How do we know that the ring of material is out there? Astronomers have suspected its presence from the first few months after the supernova explosion. The telltale evidence first appeared in spectra taken with the International Ultraviolet Explorer (IUE), a satellite operated jointly by NASA and the European Space Agency. In late May and early June of 1987, these spectra began to show the presence of sharp lines due to the presence of nitrogen, carbon, oxygen, and helium atoms. The spectra also revealed the velocity of this gas, and the motion was far too slow to be coming from material blasted out by the supernova itself. Instead, speculated astronomers, it was gas from a shell surrounding the supernova, lit up briefly by the intense burst of light from the explosion at its center.

The discovery of this circumstellar gas provides a tidy explanation for the mysterious rebrightenings of past supernovae as well as the expectation of more activity from Supernova 1987A. But in 1987 it was impossible to do much more than speculate on what was to come. The circumstellar gas was a phantom presence, otherwise unseen. The glare from the brilliant supernova made it virtually impossible to get pictures of anything in the vicinity for several years. By 1990, however, the supernova faded enough that astronomers finally were able to take high-resolution photographs of the circumstellar gas. A few photos from earthbound telescopes captured light from this gas, but the most striking photo came from the Hubble Space Telescope in late 1990. Taken through a filter that isolates a strong spectral line produced by oxygen gas, the HST photo shows a ring of material about 1.8 arc-

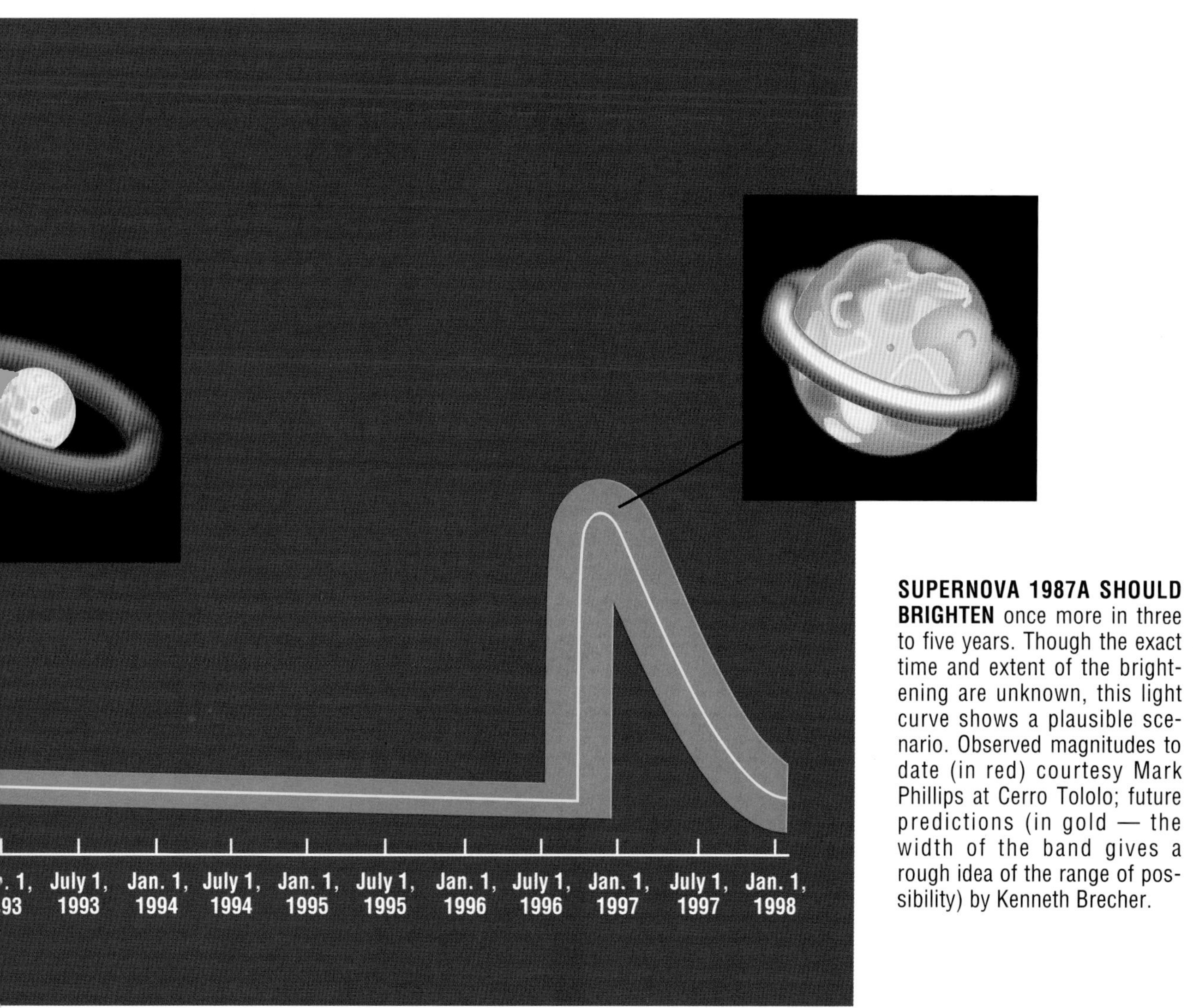

SUPERNOVA 1987A SHOULD BRIGHTEN once more in three to five years. Though the exact time and extent of the brightening are unknown, this light curve shows a plausible scenario. Observed magnitudes to date (in red) courtesy Mark Phillips at Cerro Tololo; future predictions (in gold — the width of the band gives a rough idea of the range of possibility) by Kenneth Brecher.

seconds in diameter. Because the supernova lies approximately 170,000 light-years from Earth, that angular diameter places the ring a bit less than 1 light-year from the supernova. With the supernova's debris rushing outward at about one-tenth the speed of light, the first shreds of debris should hit the ring a bit less than ten years after the explosion, or sometime in the mid-1990s.

The collision could come somewhat earlier or later than that because the Hubble image shows only the brightest part of the ring. Astronomers don't know for certain how thick the ring really is, or how far in or out it extends. The fastest-moving shreds of debris from the supernova may reach the innermost gas in the ring several years before they get to the ring's densest parts. As a result, the collision may take place over a long period of time, more like the slow-motion crumpling of two colliding trains than a sudden explosive impact. The first signs of brightening could begin as early as 1995 and extend for a year or more.

We don't know for sure how the collision will appear from Earth, but it clearly will not be as spectacular as the original blast. As the outrushing debris plows into the surrounding ring, the gas will heat up, and the more abrupt the collision, the hotter it will get. The collision will generate radio waves and X rays, which will rise in intensity for a while and then gradually decrease as the clouds slow down and cool off. Dust grains in the debris will heat up and glow, releasing infrared radiation. And hydrogen atoms will be shocked into luminescence, causing the region to brighten visibly with hydrogen's characteristic red glow.

To observers, then, the reborn supernova will appear as a tiny dot of light, as bright as 9th magnitude if the collision is particularly abrupt and intense. Though the entire ring of circumstellar gas may be glowing, it will appear about an arc-second in diameter, and only telescopes like the Hubble will be able to discern any detail. Though not as bright as it was during its brief glory days, it will once more stand out from its neighbors. It will be a fitting last hurrah for a supernova that has given us so much to observe already. □

Laurence A. Marschall is a professor of physics at Gettysburg College in Gettysburg, Pennsylvania. His last article for ASTRONOMY was "Supernova Aftermath" — about the Cassiopeia A supernova remnant — which appeared in the February 1989 issue. Kenneth Brecher is a professor of astronomy and physics at Boston University in Boston, Massachusetts and director of the Boston University Science and Mathematics Education Center.

STAR DUST

Wispy clouds of cosmic dust permeate the space between the stars.

by Gerrit L. Verschuur

Astronomers have long sought to understand the dark clouds that loom among the stars in the Milky Way. For more than half a century, they have known that stars are born from interstellar matter. But they still do not know what happens in the earliest stages of star formation, before matter coalesces into the blobs that give birth to stars. By studying the dark clouds in the Milky Way and learning their most intimate secrets, astronomers hope to shed light on this fascinating chain of events.

The launch of the Infrared Astronomical Satellite (IRAS) in 1983 set the stage. The satellite provided the first clear and comprehensive view of the infrared universe from above Earth's atmosphere, which blocks most infrared radiation. Astronomers from around the world have been milking the IRAS database ever since to learn as much as they can about the interstellar medium.

The story begins in 1984, when a team of IRAS astronomers led by Frank Low of Caltech announced the discovery of a new component of the interstellar medium. The IRAS instruments revealed that infrared radiation with a wavelength of 100 microns emanated from much of the sky. At first the astronomers suspected something was wrong with the equipment, but after checking everything thoroughly, they felt confident that what they had found was real. The 100-micron images showed wispy structures that reminded the team of cirrus clouds on Earth, so this stuff quickly came to be called interstellar "cirrus." The structures spanned huge volumes of space, ranging from light-years to many tens of light-years across.

The next step — taking a good look at these structures — posed unexpected and time-consuming challenges. The discovery of cirrus was one thing. Producing clear images of what it looked like was another story entirely.

The essence of the problem was that the IRAS designers never intended the satellite to create images. Yet close study of something as widespread as cirrus required pictures to show the full extent of their structure. These pictures could then be compared

ORION COMES ALIVE to the sensitive infrared eye of IRAS. Dominating the image are the Orion Nebula at lower right and the Rosette Nebula at middle left. Just above the Orion Nebula lies the nebulosity surrounding Zeta Orionis, and the large ring at upper right is the remnant of an ancient supernova explosion centered near Lambda Orionis. Image processing by IPAC under contract to JPL.

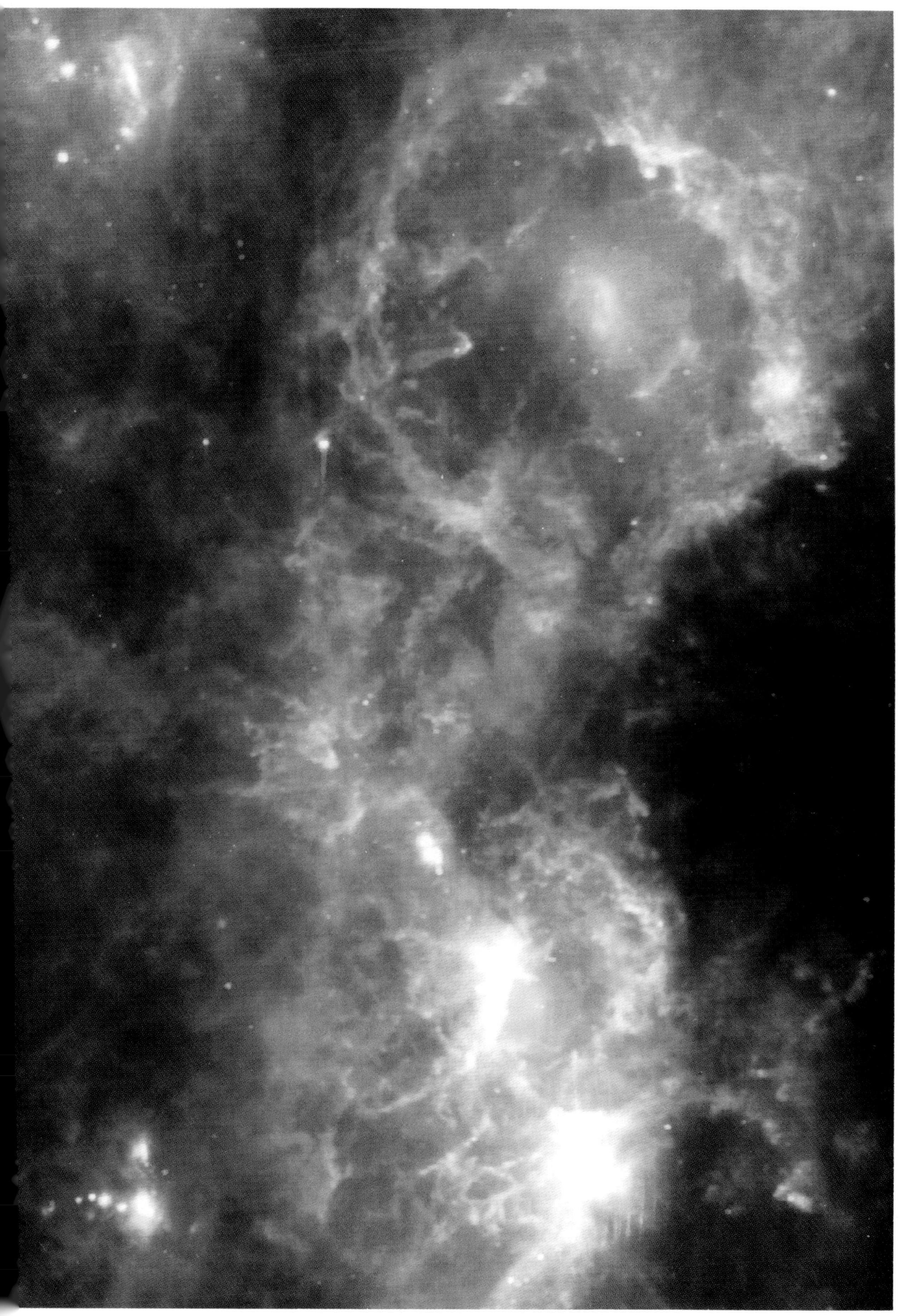

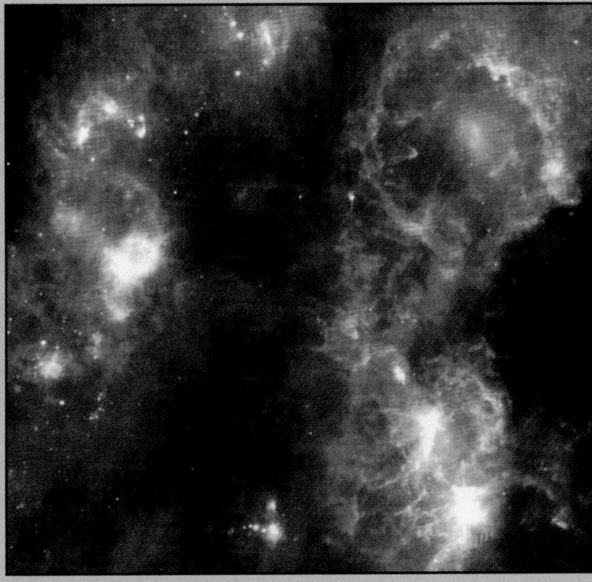

DRAMATICALLY DIFFERENT VIEWS of Orion arise depending on how you observe it. IRAS's 100-micron observations zero in on the cool, infrared-bright dust and record many wispy structures called cirrus (above). An optical photo (left), on the other hand, records bright stars and hot gas. Photo by Michael Stecker.

with optical photographs and radio images of the distribution of interstellar hydrogen in the same direction as the cirrus. This would yield a better understanding of the relationship among the various components of the interstellar medium.

Undaunted, dozens of astronomers accepted the challenge and labored for years to produce pictures of the infrared cirrus. Some of their results are shown for the first time on these pages. This small gallery of cirrus images will give you an inkling of the beauty present in the infrared sky. But it is what the images reveal about the interstellar medium, the birthplace of stars, that drove astronomers to create the images in the first place.

Weaving an Infrared Picture

It took many years to produce these images because of the way IRAS observed the sky. Unlike an imaging system, in which the telescope points to a specific area and then makes an exposure, the IRAS telescope scanned across the sky day after day. This produced a huge raster pattern of the intensity of infrared emission in four wavelength bands, at 12, 25, 60, and 100 microns. Though these data were ideal for measuring the brightness of infrared-bright stars or distant galaxies, they were terrible for constructing images. Creating an image required weaving together scans taken hours or even days apart. Because individual scans had slightly different background levels due to instrumental effects, stripes appeared across the images. Each change in background level had to be identified, measured, and then removed.

Destriping, as the process came to be called, taxed the ingenuity of even the brightest computer programmers. Learning how to do it correctly took years of effort. It actually became something of a fetish among those working with the IRAS data to be able to destripe better than the next person, and arguments broke out over whose computer programs did the best job. At the same time, astronomers at the Infrared Processing and Analysis Center at Caltech were systematically destriping all the IRAS data. That work is only now nearing completion.

Another problem faced those who wanted to use the 100-micron data from IRAS to make pictures of the cirrus: dust in the solar system. This dust not only reflects sunlight, producing a subtle glow visible to the human eye and known as the zodiacal light, but also absorbs sunlight and then re-emits that energy in the infrared. The glow forms a great swath across the heavens that follows the ecliptic, the apparent path of the planets. This local infrared glow forms a haze through which the infrared radiation from the more distant Milky Way has to shine. Only after subtracting the zodiacal light could astronomers clearly see the cirrus.

Once astronomers overcame all the problems, the results proved spectacular. After destriping and removing the zodiacal contribution, the 100-micron data provided a beautiful new view of interstellar matter. Now dozens of astronomers around the world are concentrating on their favorite regions, trying to explain why the cirrus looks the way it does and how it is associated with hydrogen gas, molecules, magnetic fields, and the structure of the Milky Way.

Clouds of Darkness

Before IRAS discovered the cirrus, dark dust clouds between the stars had been observed only with optical telescopes. One of the best known of these clouds is in the constellation Ophiuchus, where

HOLES IN THE HEAVENS — so thought some early observers of the dust clouds that block distant starlight. An optical photo of Ophiuchus (above) reveals two such dark streamers trailing to the lower left. Photo by Gary Emerson. A 100-micron IRAS image of the same area (right) shows these streamers as bright filaments. Image courtesy Eugene de Geus.

dark lanes of obscuring material block some of the background stars from view. A few clouds can even be spotted with the naked eye, such as the dark lanes in Cygnus and the almost black void near the Southern Cross known as the Coal Sack.

Today it seems almost quaint that for more than a century the dark patches between the stars were believed to be "holes" in the heavens. That notion goes back to Sir William Herschel, who concluded in the late eighteenth century that the holes allowed observers to see into the dark void beyond the stars. In 1910 Edward E. Barnard, pioneer photographer of these dark patches, told of the experiences of an astronomer who one night aimed his small telescope a little north of Antares. "Presently no stars came into the field of his telescope. After watching for some time he finally concluded the sky had clouded over, but on looking out he found it perfectly clear. His telescope had been pointed to this lane (near the star Rho Ophiuchi) and nothing but blank sky had passed." This illustrates how dark these clouds can be.

These clouds are in fact vast, swirling masses of tiny interstellar dust grains made of silicon and carbon and surrounded by water ice laced with organic molecules. These include ethyl and methyl alcohol, formaldehyde, and cyano-acetylene. The grains occupy only a small fraction of the space in a dark cloud, but the clouds are so huge that collectively the tiny particles block starlight from beyond. Thus we see the clouds as dark markings against a bright backdrop of stars in the Milky Way.

The photos above compare the 100-micron counterpart of the dark clouds in Ophiuchus to an optical view. The bright infrared streamers radiating away from the core of Ophiuchus correspond to the dark dust lanes that intrigued Barnard. Within these filaments, astronomers have observed the infrared glow of low-mass stars in the process of being born.

The apparent difference between the dark dust lanes seen in the optical and the bright streamers seen in the infrared arises from the temperature of the dust. Everything in the universe emits some form of electromagnetic radiation. That includes your body which, at a temperature of 295 kelvins (about 20° Celsius), glows brightly in the infrared. That is why infrared-sensitive binoculars can be used to reveal the presence of enemy soldiers on an otherwise pitch black night.

Like all forms of electromagnetic radiation, infrared can be generated by the motion of the particles that make up an object. Because an object's temperature is above absolute zero, motion at the atomic level is always present and the object will emit some radiation at all wavelengths. However, as an object grows hotter, the wavelength at which it emits the most energy becomes shorter. In the case of the Sun's surface, which glows at a temperature of 5,800 kelvins (K), the radiation peaks in the visible part of the spectrum. Interstellar grains, on the other hand, are warmed by starlight to only about 20 or 30 K. At this temperature, they are expected to radiate infrared energy at a wavelength of around 100 microns, and that's exactly what IRAS found.

Blanketing the Sky

No one was surprised that IRAS detected the glow of interstellar dust, at least not the emission coming from the plane of the Milky Way. What came as a shock was that the cirrus appeared all

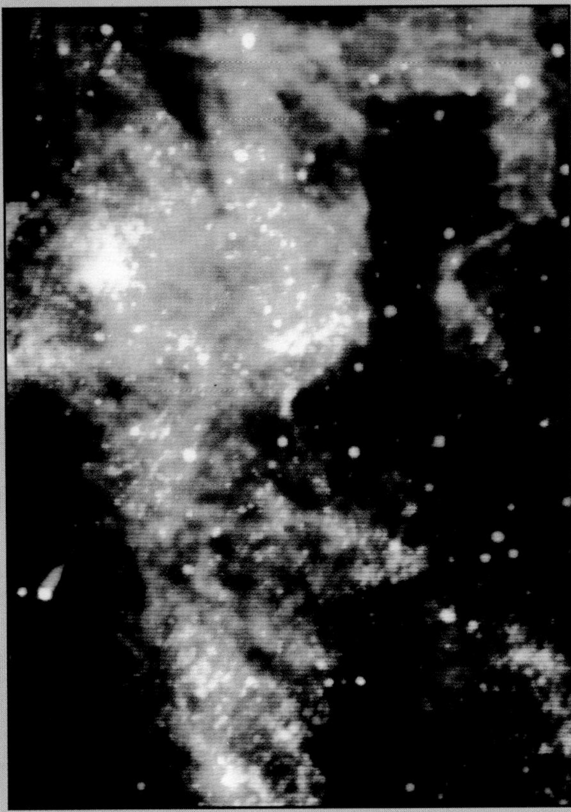

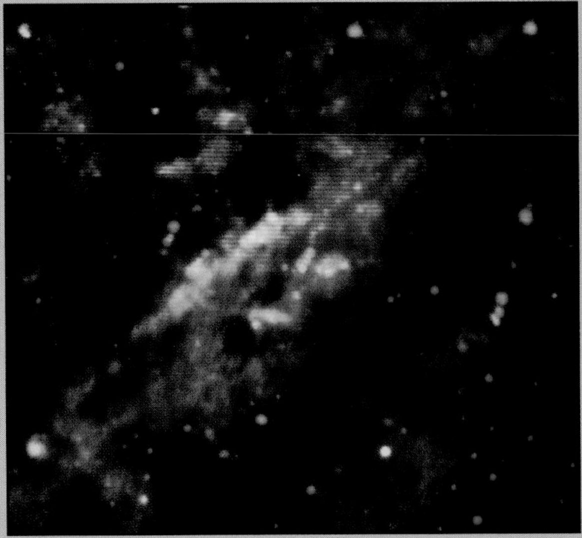

THE BACKDOOR NEBULA is a cirrus structure with a distinctly rectangular shape (at upper right in photo at left). The bright area to the nebula's left is home to three carbon monoxide-rich clouds. Image courtesy Lee J. Rickard, Fran Verter, and Barry Turner. Cirrus filaments abound in the Eridanus region (above). Image courtesy Lee J. Rickard.

over the sky. Though interstellar dust is present in small quantities everywhere in the Galaxy, before the age of IRAS, it could only be seen against the bright background of stars in the Milky Way. Diffuse stretches of this material well away from the Milky Way, where the number of background stars is much lower, essentially remained invisible, even in long exposure photographs.

In displaying the IRAS data, astronomers typically have assigned each of the wavelength channels a different color and then combined them into one image. In most cases, the long wavelength, 100-micron emission is colored red while the short wavelength, 12-micron emission is made blue. Green is used for the 60-micron emission. A blue dot in an image is usually a star that shines brightly in the infrared, often because it is surrounded by dust at a temperature of a few hundred kelvins. A slightly green spot is likely to be a galaxy (galaxies tend to glow brightest in the infrared at 60 microns). The wispy red structure in the multicolor images is the cirrus. Depending on its temperature, the cirrus may produce relatively more 12-micron emission and then appear white in the image.

The centerpiece of the photo above is a cirrus structure called the Backdoor Nebula. Its rectangular shape and nearly straight sides have intrigued Lee J. Rickard of the Naval Research Laboratory, Fran Verter of the Goddard Space Flight Center, and Barry Turner of the National Radio Astronomy Observatory. The appearance of the structure is purely fortuitous; still, they have been heard to joke that this doorway may offer us a view into another dimension.

The bright area to the left of the Backdoor Nebula contains three carbon monoxide-rich "clouds." First spotted by University of Maryland graduate student Loris Magnani, these clouds are better understood today because of the satellite's observations. Before IRAS began its survey, Magnani took a close look at the Palomar Sky Survey plates, searching for and finding many previously unrecognized and very faint veilings of dust cutting off light from distant stars. Next he and his collaborators Leo Blitz and Lee Mundy looked in those directions for the radio signature of carbon monoxide molecules. Their radio survey turned up carbon monoxide virtually everywhere they looked, in all of the 50 or so previously unnoticed clouds. But it was the 100-micron data from IRAS that revealed the nature of these structures in glorious detail.

Because carbon monoxide is not a particularly stable molecule, it gets destroyed easily when exposed to starlight. Its presence, therefore, indicates that something is shielding it from starlight, namely, interstellar dust. IRAS confirmed the existence of this dust when it found the 100-micron emission from cirrus associated with the carbon monoxide structures observed by Magnani, Blitz, and Mundy. Since then radio astronomers have found in these structures other molecular species, such as ammonia and formaldehyde.

Rickard and his colleagues also detected weak formaldehyde absorption at the Nebula's edge, in areas of apparently low density. Because formaldehyde can only absorb radio signals when it is gathered in very dense clumps, this discovery meant that the cirrus is filled with tiny, dense patches that are invisible in the 100-micron images. Some astronomers think these clumps may represent the very earliest stages of star formation, when interstellar gas and dust are just beginning to swirl around. Gravity would then begin

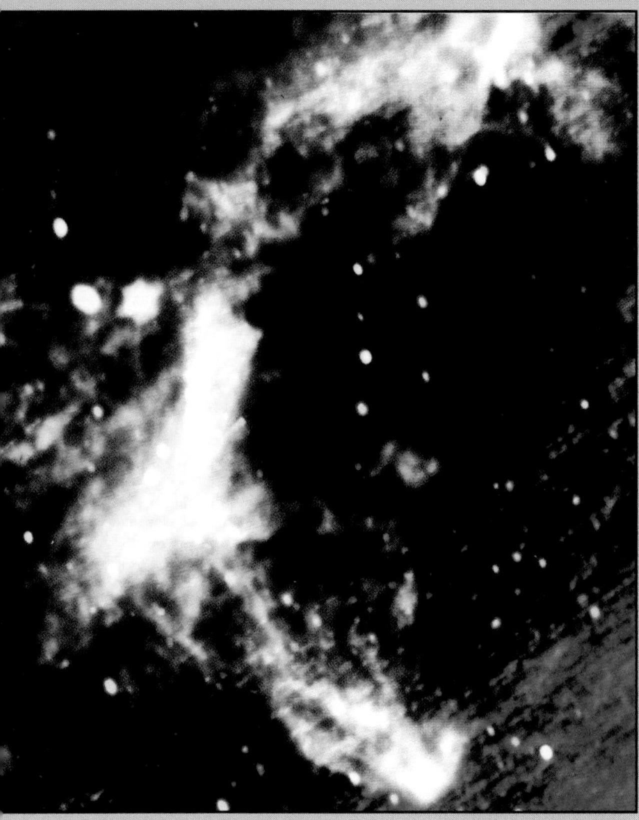

A RING OF CIRRUS dominates part of Ursa Major (left). The brightest emission in this IRAS image coincides with clouds of carbon monoxide. Courtesy F. Desert and F. Boulanger. The galactic plane is defined by the prominent cirrus structure seen in this view of the 100-micron emission from about one-third of the Milky Way Galaxy (below). Courtesy Lee J. Rickard, who used the data file produced by F. Boulanger.

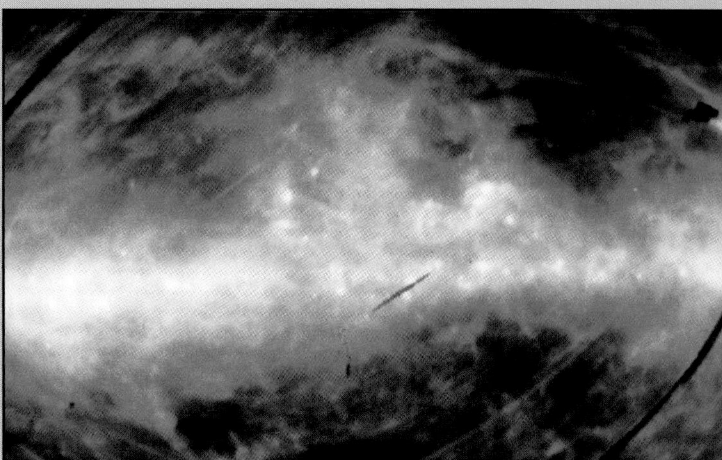

to pull on the gas and dust so as to draw the matter together to the point where star formation can be initiated, perhaps a few million years hence.

Among the photos above is a particularly striking one that shows cirrus filaments at the southern end of a huge ring of dust and gas (about 35° in diameter) called the Eridanus shell. At its northern end, Barnard's Loop, just northeast of the Orion Nebula, defines its boundary. These cirrus filaments are associated with both cold (50 K) and hot (10,000 K) hydrogen gas filaments, the former seen with radio telescopes and the latter with optical telescopes. This complex shell structure apparently was created long ago by the combined action of supernovae in the star-forming region in Orion. Strong ultraviolet radiation streaming out of the young stars in Orion keeps this vast cavity excavated and some of the interstellar matter is pushed into filaments along its outside edge.

Down to the Nitty-Gritty

Just what is the cirrus dust made of? French astronomers F. X. Desert, F. Boulanger, and J. L. Puget have used the IRAS data in combination with theoretical models of how dust obscures starlight to conclude that a typical cirrus structure contains grains of at least three distinctly different sizes. Most of the grains are relatively large, between 10 and 100 millionths of a millimeter across, though still tiny by human standards. The rest is in the form of even smaller grains (between 1 and 10 millionths of a millimeter in size) and polycyclic aromatic hydrocarbons, giant molecules based on carbon chains. Each of these molecules contains hundreds of atoms and is less than one millionth of a millimeter in diameter. The bigger grains are made of silicates surrounded by water ice containing various metals and complex molecules. The smaller grains are carbon-rich.

Individual astronomers continue to use the IRAS database to extract the information they need to make images of their favorite regions. At the same time, astronomers at the IRAS data analysis centers in the United States and Europe are working to complete the task of systematically removing the zodiacal light and destriping the data. Within a year or so, astronomers will be able to study the 100-micron cirrus as readily as they might photograph an area of sky with a telescope.

The main difference between photographing the visible sky and studying the infrared heavens will be that the infrared sky resides inside a computer, which will allow you to produce images on a video monitor. Those images will be like those shown here, and they can be analyzed to provide information on the temperature, density, and structure of the dust. This information will pave the way for a comprehensive understanding of how matter between the stars gathers into blobs that will someday form stars.

The saga of producing images from the IRAS data has been one of overcoming obstacles. Astronomers from many nations have collaborated to figure out how to destripe the data and remove the zodiacal emission to produce images undreamed of when the satellite was launched. Their work gives us a wonderful new view of what matter between the stars really looks like. □

Gerrit L. Verschuur is a radio astronomer and author living near Bowie, Maryland. His last article for ASTRONOMY *was "The End of Civilization" in the September 1991 issue.*

The Best BLACK HOLE in the GALAXY

Move over, Cygnus X-1 — astronomers now believe a distant binary system in the constellation Monoceros is the best case yet for a stellar black hole.

by Ken Croswell

AN ORANGE DWARF ORBITS about the center of the A0620-00 binary system in Monoceros, indicating that its dark companion is a black hole of at least 3.8 solar masses. Painting by Joe Bergeron.

On Earth, black holes are everywhere. The national debt is a black hole. The savings-and-loan mess is a black hole. A friend of mine recently claimed his graduate thesis was also a black hole. "It sucks up all my time, all my energy, all my happiness," he said. "And there's no escaping it."

In our Galaxy, however, black holes are much harder to find. For most of the last twenty years, astronomers had good evidence for only one stellar black hole in the entire Galaxy: a source of X rays in the constellation Cygnus called Cygnus X-1.

Now, though, astronomers have uncovered a much better candidate for a black hole in our Galaxy. It lies in the constellation Monoceros some three or four thousand light-years away and bears the prosaic name A0620–00. The A refers to Ariel V, the X-ray satellite that first detected the object, and the numbers refer to the object's right ascension and declination. A0620–00 was discovered in 1975, when it emitted a shower of light and X rays. Observations soon revealed that A0620–00 was a binary consisting of an orange dwarf star and a dark companion. Astronomers continue to observe the object and recent studies of the orange dwarf's motion around the dark star indicate that the dark star must be a black hole. Despite its less appealing name, A0620–00 is a better candidate for a black hole than Cygnus X-1.

Black-hole hunters look for dark objects whose mass exceeds three solar masses.

Light-Devouring Holes in Space

A black hole is an object with so much gravity that nothing can escape it — not even light, the fastest thing in the universe. Anything approaching a black hole gets pulled into the object and disappears as if it fell into a hole. Because even light cannot escape, the hole appears black.

Gravity is the key to a black hole's immense power. The black hole's strong gravity keeps captured material from escaping. Of course, every planet, moon, and star has gravity. Earth has enough gravity that you have to travel faster than 11 kilometers per second to overcome the force of gravity. This number is Earth's escape velocity. The gravity of Jupiter is even stronger: its escape velocity is 60 kms per second. A black hole has so much gravity that to escape one you would have to travel faster than the speed of light, 300,000 kms per second. But traveling faster than light is impossible, so once you get into a black hole you can't get out.

The gravitational force you feel on the surface of a planet depends on both the planet's mass and its radius. The greater the mass, the stronger the gravity and the greater the escape velocity. Less obviously, the smaller the radius, the stronger the gravity and the greater the escape velocity, too. The reason: you are then standing closer to the planet's center, so the gravitational pull on you is stronger and you have to travel faster to escape.

Anything can become a black hole if you compress it enough. For example, if Earth were the same mass it is now but had only one-fourth its present radius, the escape velocity of someone standing on its surface would be twice what it is now. If you compressed Earth still further, the escape velocity would get still higher. If Earth's diameter shrank to less than one centimeter, the escape velocity would exceed the speed of light.

Earth, of course, will never become a black hole, nor will the Sun. When the Sun runs out of fuel, it will collapse and form a white dwarf. But when a very massive star runs out of fuel, it collapses and forms a black hole.

Like the Sun, most stars support their huge weights by generating energy. This energy keeps the stars warm, and the pressure of the warm gas holds the stars up against their own gravity. Someday, though, a star exhausts its fuel. Without its energy source, the star can no longer hold itself up against the pull of its own gravity and so the star collapses. A star with an initial mass below eight solar masses collapses into a white dwarf star. White dwarfs hold themselves up not through warmth but through the pressure exerted by the star's electrons on each other.

There is a limit to how much pressure electrons can exert, however, and that limit is exceeded if the white dwarf's mass is above 1.4 solar masses. White dwarfs more massive than 1.4 solar masses therefore cannot exist. But all stars with initial masses below eight solar masses lose so much mass during their lives that they form white dwarfs with masses under 1.4 solar masses. In fact, the average mass of a white dwarf is only 0.6 solar mass.

What happens if a star is born with more than eight solar masses? After the star runs out of fuel, the outer portion explodes, forcing the core of the star to collapse. The remaining star exceeds 1.4 solar masses and so cannot form a white dwarf. Instead, as the star collapses, its electrons and protons smash together to form neutrons, and a neutron star is born. Whereas the mutual repulsion between its electrons supports a white dwarf, the mutual repulsion between neutrons supports a neutron star.

Astronomers have catalogued hundreds of neutron stars throughout the Galaxy. Some neutron stars make their presence known because they are pulsars, emitting periodic pulses of radio waves and/or X rays.

As with white dwarfs, there is a limit to how much mass a neutron star can have. If the collapsing star exceeds two or three solar masses (the exact figure is unknown, but depends upon the interaction between the neutrons), the star's gravity overwhelms the support the neutrons provide. In a fraction of a second, the star shrinks. As the star gets smaller, its surface gravity gets stronger and the escape velocity climbs higher until it exceeds the speed of light. The star has now become a black hole.

Which stars form black holes? Again, astronomers do not know the precise answer. Most stars with initial masses above eight solar masses probably form neutron stars rather than black holes, but the most massive stars of all — stars born with masses above about 40 Suns — probably form black holes.

How to Find a Black Hole

Because neutron stars with masses above three solar masses do not exist, your goal as a black-hole hunter is to find a dark object whose mass exceeds three solar masses. Then you know you have a black hole.

But finding a black hole is tough for two reasons. First, black holes obviously radiate no light. Second, the stars that eventually become black holes are quite rare. Fewer than one star in ten million has a mass exceeding 40 solar masses. Thus, black holes must be rare. Yet, of the billions of stars in our Galaxy, several thousand of them should become black holes.

Because black holes are black, a black hole isolated in space is probably impossible to discover. But if a normal star orbits the black hole, we can detect the gravitational influence of the black hole by noticing that the normal star circles an unseen companion.

Such binary systems call attention to themselves if the black hole grabs material from the normal star. The material spirals into the black hole like water going down a drain. Before this material takes the final plunge, however, it forms a disk around the black hole called an accretion disk.

The accretion disk is extremely hot and emits X rays, which astronomers can detect. The accretion disk is hot because material is falling into it. Whenever something falls, it heats up: water falling through Hoover dam, for example, is one degree warmer at the bottom than at the top. The energy released by the falling water is converted into heat. Material falling into the accretion disk of a black hole gets heated by gravity — and by friction — and since the gravity is so strong, the temperature rises above a million degrees. Because it is so hot, the accretion disk radiates X rays. That was how Cygnus X-1 was found.

Cygnus X-1

In the 1960s, astronomers noticed X rays coming from a region in Cygnus and named the source Cygnus X-1. In 1971, they found that the X-ray source coincided with a normal 9th-magnitude star called HDE 226868. This star had been catalogued years before as a blue star of spectral type O. O-type stars have high masses and therefore are less common

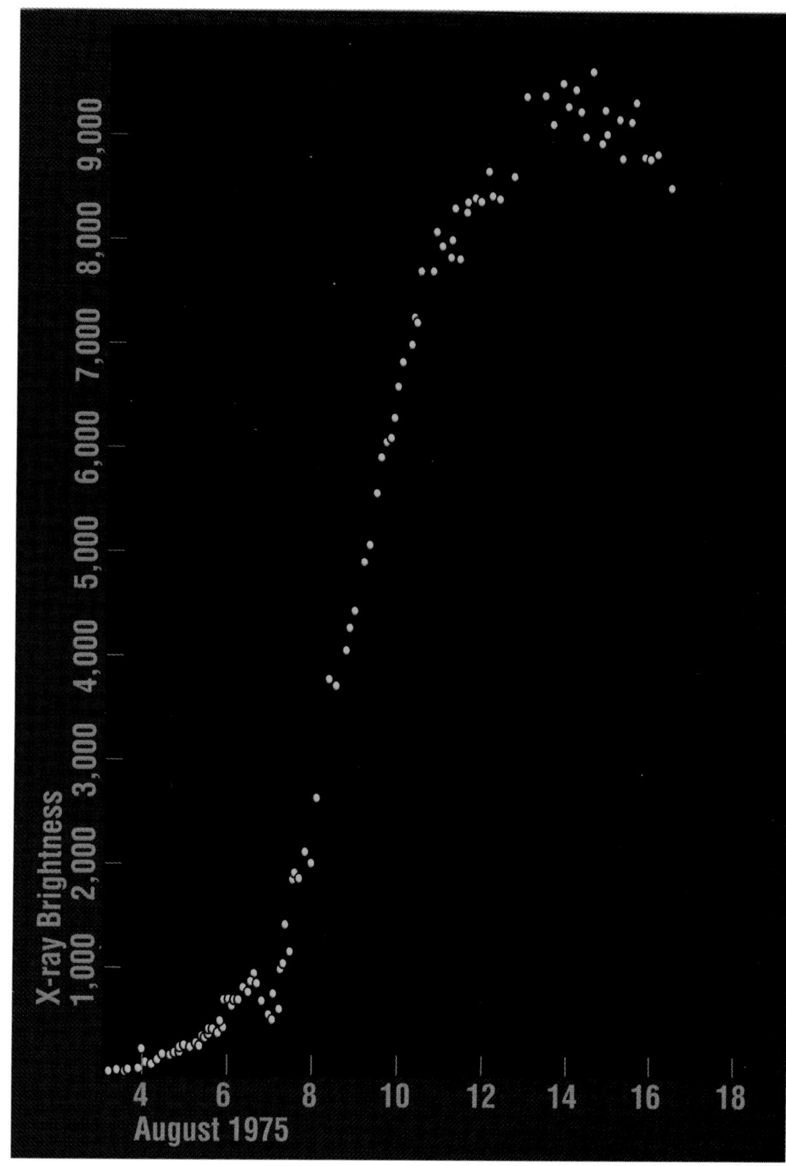

A0620–00 WAS DISCOVERED in 1975 when astronomers using the Ariel V X-ray satellite noted this X-ray outburst. An optical outburst followed the peak intensity of the X-ray transient source. Illustration from data by University of Leicester X-ray astronomy group.

than the low-mass stars that make up most of our Galaxy. Still, O stars are not that unusual and several are even visible to the naked eye.

But HDE 226868 *was* unusual. Astronomers measured the star's velocity and found that the star revolved around something every 5.6 days. But the companion of HDE 226868 turned out to be dark and invisible. By assuming the O star has a high mass, as most other O stars do, astronomers concluded that in order to perturb the O star, the dark star had to exceed three solar masses. Cygnus X-1 was likely a black hole.

But it was not a definite black hole because the dynamics of the binary system say only that the dark companion must exceed 0.25 solar masses. The argument that the dark star exceeds three solar masses depends on our assuming the O star has a high mass. If the O star has less mass than usual for its spectral type

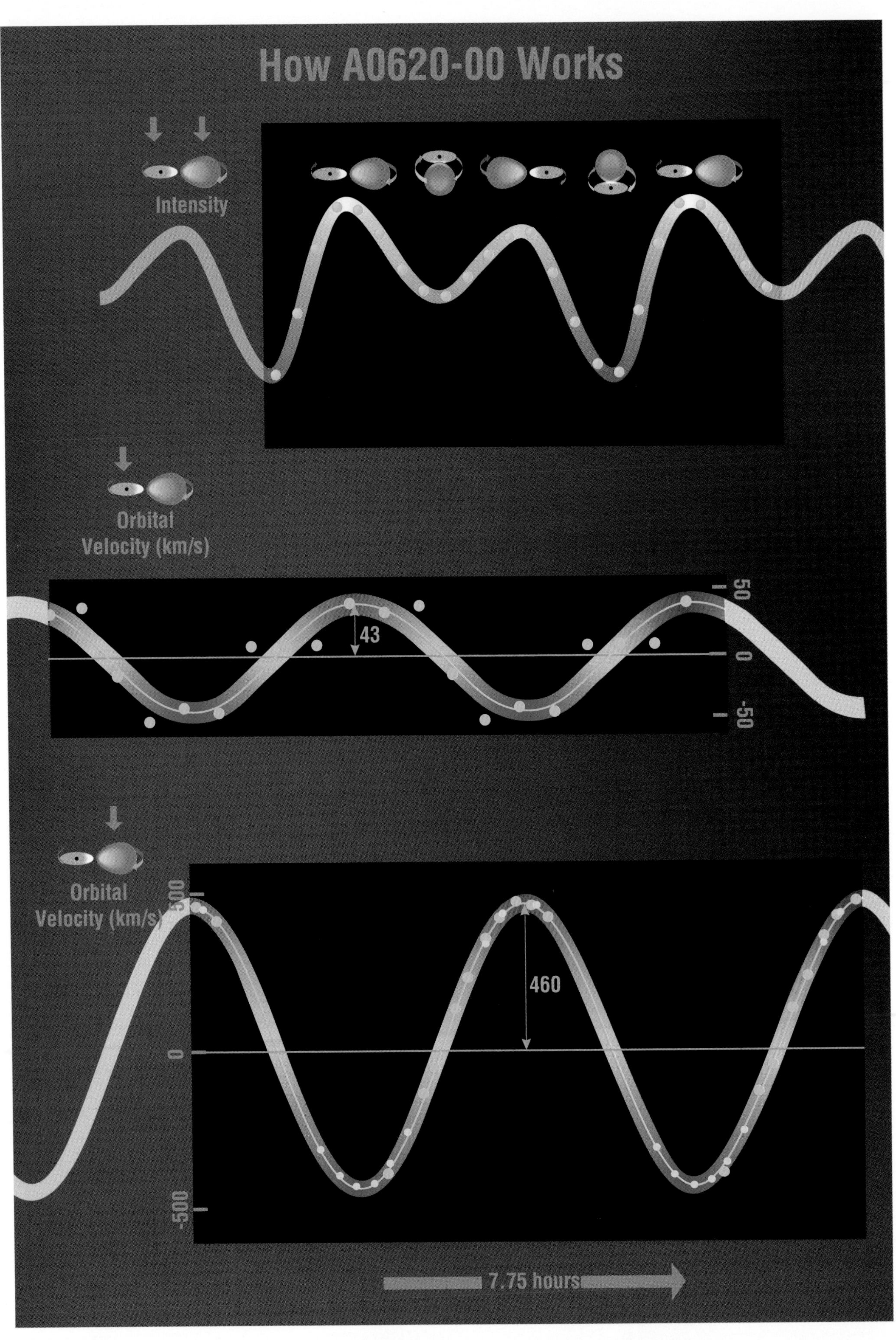
How A0620-00 Works
Intensity
Orbital
Velocity (km/s)
50
0
-50
43
Orbital
Velocity (km/s)
500
0
-500
460
7.75 hours

— as do many stars in X-ray binary systems — then the unseen companion could contain less than three solar masses. In fact, some astronomers have constructed models of Cygnus X-1 in which the dark companion is only a neutron star rather than a black hole.

When dealing with exotic objects like black holes, astronomers want absolute proof. Cygnus X-1 *probably* is a black hole. But the dark star could have less than three solar masses, so Cygnus X-1 does not give us the proof we desire. Now, though, astronomers have found a perfect black hole, a dark star whose mass definitely exceeds three solar masses: A0620–00.

An X-ray Outburst in Monoceros

On August 3, 1975, the British satellite Ariel V detected a source of X rays in the constellation Monoceros, near its border with Orion. The source, A0620–00, rapidly intensified: on August 5, it surpassed the strength of the Crab Nebula; shortly after, it exceeded Cygnus X-1; and on August 8, it outdid Scorpius X-1, the strongest X-ray source in the night sky. For nearly two months, the object remained brighter than any other X-ray source except the Sun.

Then astronomers discovered an 11th-magnitude blue star at the position of A0620–00. Photographs taken at Palomar Observatory in 1955 indicated the star was normally reddish and 18th magnitude. So the X-ray outburst coincided with an optical nova, which was named V616 Monocerotis.

Searching astronomical archives, astronomers found that the nova had erupted before. In November 1917 it attained 12th magnitude, but no one had noticed it. Later examination of old sky patrol plates at Harvard University revealed the nova's outburst.

The light and X rays of A0620–00 eventually faded. Today an 18th-magnitude orange star lies in this place. The star is a dwarf of spectral type K5, somewhat fainter, cooler, and smaller than the Sun. These stars are common (about 15 percent of all stars in our Galaxy) so we know a lot about them. If the orange dwarf in the A0620–00 system is a normal K5 dwarf, it has a mass around 0.7 solar mass and an intrinsic brightness five to ten percent that of the Sun. Since its apparent magnitude is 18, astronomers estimate the distance to this star to be roughly three or four thousand light-years. But A0620–00 is not a normal orange dwarf. It circles a black hole.

A Perfect Black Hole

As the X-ray outburst proved, there was more to the A0620–00 system than just an orange dwarf. After the outburst faded, observations revealed that the K star contributes only about half the total visible light the system emits. The other half comes from a hot accretion disk surrounding the orange star's unseen partner.

In 1983, Jeff McClintock of the Harvard-Smithsonian Center for Astrophysics and colleagues reported that the orange dwarf's brightness varies 0.2 magnitude every 7.75 hours, which indicates the star revolves around something every 7.75 hours. Twice during each revolution, the star's brightness, as observed on Earth, reaches a maximum; twice it reaches a minimum.

McClintock and colleagues suggested the orange dwarf varies in apparent brightness because it is not spherical. Instead, the intense gravity of the dark companion distorts the orange dwarf into an egg shape. As the orange dwarf orbits around its companion, the same side points toward the dark star. Thus, the dwarf presents different profiles to us, varying the brightness as seen from Earth (see diagram on opposite page). Twice during each revolution, we view the egg-shaped star sideways and its apparent brightness attains a maximum; twice during each revolution, we view the star end-on and its apparent brightness reaches a minimum.

A0620-00 is now the best black-hole candidate.

What was the orange dwarf's companion? A black hole or just another neutron star? To find out, astronomers had to determine the mass of the dark star by studying the orange dwarf's orbit. If the dark star's mass exceeds three times the Sun's, the companion must be a black hole.

Unfortunately, the orbital period of the binary system as found from brightness variations does not reveal by itself the mass of the dark companion. Astronomers also need to know the separation between the dark star and the orange dwarf. For a given period, the greater the separation, the greater the mass must be. For example, if a star has a planet circling it yearly and if the planet lies one astronomical unit — the distance from the Sun to Earth — from the star, then the star must have exactly the same mass as the Sun. If, on the other hand, a planet completes an orbit once a year but lies two astronomical units from the star, the star must be eight times more massive than the Sun.

Two stars in a binary like A0620–00 actually circle their center of mass, which lies between them, instead of orbiting one another. The binary's center of mass is the place where you would put a fulcrum to balance a giant seesaw holding one star at each end. The center of mass, therefore, always lies closer to the more massive star. For example, if one star has ten times the mass of the other, the center of mass lies a tenth as far from the high-mass star as from the low-mass star. In the case of A0620–00, the center of mass lies closer to the dark star than to the orange dwarf because the dark star has more mass. Every 7.75 hours, the orange dwarf revolves about this center of mass and so does the dark star. But because the center of mass lies closer to the dark star, the dark star does not move as much as the orange dwarf does. The total separation between the dark star and the

THE BRIGHTNESS OF THE BINARY SYSTEM (top) changes as the stars orbit each other every 7.75 hours. Orbital velocities of the black hole (middle) and orange star (bottom) change between negative and positive as the two approach and recede. Dots show actual data for the system. Illustration by Steve Davis.

Battle of the Black Holes

Cygnus X-1

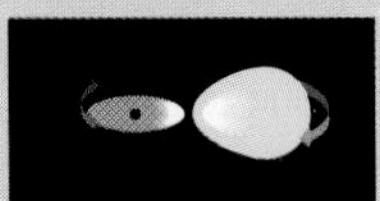

System data
distance = 8,200 light-years
period = 5.6 days
best estimate of masses
- 09.7 supergiant — 33 solar masses
- black hole — 16 solar masses

Pro
1) optical variability suggests dark star has greater than 7 solar masses
2) If O star is normal, dark star has 16 solar masses

Con
1) O star is possibly less massive than normal O stars; dark star could be a neutron star
2) distance to the binary system not well known; if closer than 5,600 light-years, dark star could be a neutron star
3) minimum possible mass for dark star — 0.25 solar mass

Large Magellanic Cloud X-3

System data
distance = 165,000 light-years
period = 1.7 days
best estimate of masses
- B3 main sequence star — 6 solar masses
- black hole — 9 solar masses

Pro
distance well known so brightness estimates can be used to place dark star mass at greater than 6 solar masses

Con
1) minimum possible mass suggests dark star close to neutron star limit
2) optical identification of B-star companion is not certain so orbital information is uncertain
3) brightness estimates do not include accretion disk emission, so dark star could be a neutron star

A0620-00

System data
distance=3,200 light- years
period= 0.32 day
best estimate of mass
- K5 main sequence star — 0.7 solar mass
- black hole — 8 solar masses

Pro
1) orbital dynamics indicate dark star must be a black hole
2) orbital parameters are well established compared to parameters of other candidates

Con
None at this time

orange dwarf — which will tell us the dark star's mass — is the same as the distance of the dark star from the center of mass plus the distance of the orange dwarf from the center of mass.

But we cannot see the separation between the two objects as we could if they were a nearby double star. Nevertheless, in 1986, McClintock and Ronald Remillard (Massachusetts Institute of Technology) put a lower limit on the separation between the dark star and the orange dwarf, which in turn puts a lower limit on the dark star's mass: 3.2 solar masses, suggestions that the dark star was indeed a black hole.

McClintock and Remillard achieved this feat by measuring the line-of-sight velocity of the orange dwarf. As the orange dwarf revolves about the center of mass, it sometimes approaches us and its light is blueshifted and, at other times, recedes from us and its light is redshifted. McClintock and Remillard measured these blueshifts and redshifts to find that the line-of-sight velocity, like the brightness, varies every 7.75 hours, confirming that the star completes an orbit every 7.75 hours. Furthermore, McClintock and Remillard found that the maximum velocity the orange dwarf attains is huge: 460 kms per second. For comparison, the Earth revolves around the Sun at only 30 kms per second.

The orange dwarf's velocity and the orbital period constrain the separation of the orange dwarf from the center of mass. If we view the binary edge-on, then the line-of-sight velocity and the period tell us how far the orange dwarf travels during a single orbit. This distance is simply the circumference of its orbit. The orbit's circumference then tells us the radius of the orbit — that is, the separation between the K star and the center of mass. The separation turns out to be 0.0136 AU. If the Sun had a planet 0.0136 AU from it, that planet would circle the Sun once every 13.9 hours. Because the period of the orange dwarf — 7.75 hours — is shorter than this, the mass of the dark star must be greater than that of the Sun. By working through the numbers, McClintock and Remillard found that the dark star had to contain at least 3.2 solar masses.

This is a lower limit because it assumes we view the binary edge-on. Chances are we don't. If, instead, the plane of the orange dwarf's orbit is tilted to our line of sight, then the true velocity of the orange dwarf (which we cannot observe) is even greater than the line-of-sight velocity (which we can observe). For example, in the extreme case when we observe a binary's orbit face on, we observe no blueshifts or redshifts at all, since neither star approaches or recedes from us.

Now, if the true orbital velocity of the orange dwarf is even greater than what McClintock and Remillard measured, the orange dwarf travels even farther in 7.75 hours, so its orbit must be larger and its separation from the center of mass greater than 0.0136 AU. This greater separation implies a greater mass for the unseen companion, because the orange dwarf now lies farther than we had previously assumed, yet it whips around the dark object so fast that it still completes an orbit every 7.75 hours.

McClintock and Remillard's figure of 3.2 solar masses is a lower limit in another sense as well. Recall that, for a given period, the mass depends on the total separation between the dark companion and the orange dwarf. McClintock and Remillard determined a lower limit only for the separation between the orange dwarf and the center of mass. In order to know the total separation between the dark companion and the orange dwarf, we must also know the distance of the unseen companion from the center of mass. The total separation between the dark companion and the orange dwarf is then the sum of the two.

The orbital dynamics of A0620–00 demand that it contain a black hole of at least 3.8 solar masses.

In 1990, Carole Haswell of the University of Texas at Austin and Allen Shafter of San Diego State University further strengthened the case for a black hole in A0620–00 by determining the separation of the dark companion from the center of mass. They measured the line-of-sight velocity not only of the K dwarf but also of the dark companion itself. They find the dark companion must be at least 3.8 times more massive than the Sun.

Of course, Haswell and Shafter could not observe the black hole directly. Instead, they measured the line-of-sight velocity of the radiation coming from the black hole's accretion disk. As expected, they found that the black hole completes a revolution around the center of mass every 7.75 hours. Furthermore, the accretion disk's line-of-sight velocity is exactly out of phase with the orange dwarf's line-of-sight velocity, as it should be: When the orange dwarf approaches us, the accretion disk recedes from us; when the orange dwarf recedes from us, the accretion disk approaches.

Haswell and Shafter found that the accretion disk — and thus the black hole — attains a maximum line-of-sight velocity of 43 kms per second. Since the orange dwarf's maximum line-of-sight velocity (457 kms per second) is about 11 times greater than this, the black hole must be 11 times more massive than the orange dwarf. Furthermore, the black hole's distance from the center of mass must be $^1/_{11}$ the distance of the orange dwarf from the center of mass. Since the orange dwarf lies at least 0.0136 AU from the center of mass, the black hole must lie at least 0.0013 AU from the same center of mass. The total separation between the black hole and the orange dwarf is the sum of these two numbers and, therefore, is at least 0.0149 AU. Together with the period of 7.75 hours, the total separation implies that the black hole's mass is at least 3.8 times the Sun's.

As before, we get this minimum value of 3.8 solar masses by assuming that we view the binary edge-on. We probably do not view the system edge-on, in which case the true velocities are greater than the velocities we measure, and the separations of each object from the center of mass are greater as well, implying a greater mass for the black hole.

What is the most likely mass for the black hole? A normal K5 dwarf has about 0.7 solar masses. Since the black hole is nearly 11 times more massive than the orange dwarf, the black hole probably has a mass around seven or eight times the Sun's. This estimate depends on our assuming that the K star is a normal orange dwarf and so could be wrong. Though a well-understood type of star, normal orange dwarfs do not circle black holes, nor are they shaped like eggs the way the K star in A0620–00 is.

Nevertheless, from the dynamics of the binary system alone, astronomers conclude that the dark object has at least 3.8 solar masses. It is too massive to be a neutron star. It must therefore be a black hole.

Despite its more poetic name, Cygnus X-1 cannot make such a claim. In fact, from the orbital motion of the O-type star in Cygnus X-1, all astronomers can say is that the dark companion must have a mass of at least 0.25 solar masses — not much of a constraint. Evidence that the dark star has a large mass comes from other arguments less secure than orbital dynamics. Thus, although the dark star in Cygnus X-1 is probably a black hole, the dynamics of the binary itself do not force us to conclude that it must be.

In contrast, the dark star in A0620–00 must exceed three solar masses and must be a black hole. We know this from the best method possible: the dynamics of the binary that contains the dark star. For no other system in the Galaxy is the evidence so strong and the argument so compelling. A0620–00 is by far the best black hole in the entire Galaxy — and the firmest evidence astronomers yet have that these exotic and extreme stars really exist. □

Ken Croswell received his doctorate in astronomy from Harvard University and has written for New Scientist, Star Date, *and many other publications. His last article for ASTRONOMY was "Encounter in Orion" in the January 1992 issue.*

RESEARCH NEWS

A New Look at the Crab Nebula

Radio observations and powerful supercomputers have combined to give astronomers a fresh view of this supernova remnant.

by Gerrit L. Verschuur

The Crab Nebula in Taurus has always held a special fascination for astronomers. Born of a supernova, the nebula's unique shape and easy visibility have fascinated sky observers for several centuries. The Crab's slowly expanding shell of gas and peculiar synchrotron emission have been studied by astrophysicists for decades. Now, after seven years of study, astronomers have given us a new look at an old crustacean.

Two radio astronomers at the University of Toronto have produced a detailed radio image of the Crab as part of their research into how the Crab's tentacles are moving. Michael Bietenholz and Philipp Kronberg used radio observations from the Very Large Array radio telescope in New Mexico to produce a detailed radio image. "Comparing our image with another made in 1982 gives us the most detailed look at the physical properties of a supernova remnant ever made," said Kronberg. So many bits of data were required to produce the image (59.4 megabytes to be exact) that a CRAY X-MP/24 supercomputer at the Ontario Center for Large Scale Computation had to be used to generate the images. "We used fifteen hours of number-crunching time on the Cray, much less than the 180 days an ordinary computer would require," said Kronberg.

The Crab Nebula is recognized as unique among supernova remnants. It does not have the expected shell-like structure of a remnant. If any shell exists it is filled with twisting filaments and glowing matter.

The Crab has a fascinating history. On July 4, 1054, Chinese astronomers unknowingly witnessed the moment of birth of this object when a "guest star" appeared just southeast of Zeta Tauri. It reached a peak magnitude of –3.5 and was so bright that it was visible during daylight for nearly a month. It faded from the night sky a year and a half later.

In 1731 John Bevis, an English physician, found a fuzzy nebula at the location of the guest star. No one paid much attention to Bevis' discovery until Charles Messier rediscovered it while trying to find Halley's Comet in 1758. He saw a "whitish light, elongated in the form of the light from a candle, not containing any star" in the constellation Taurus. An inveterate comet hunter, Messier found himself distracted by these fuzzy nebulae. For the benefit of other comet hunters he made a list of these nebulae to prevent confusion with comets. The first entry, known afterward as M1, was

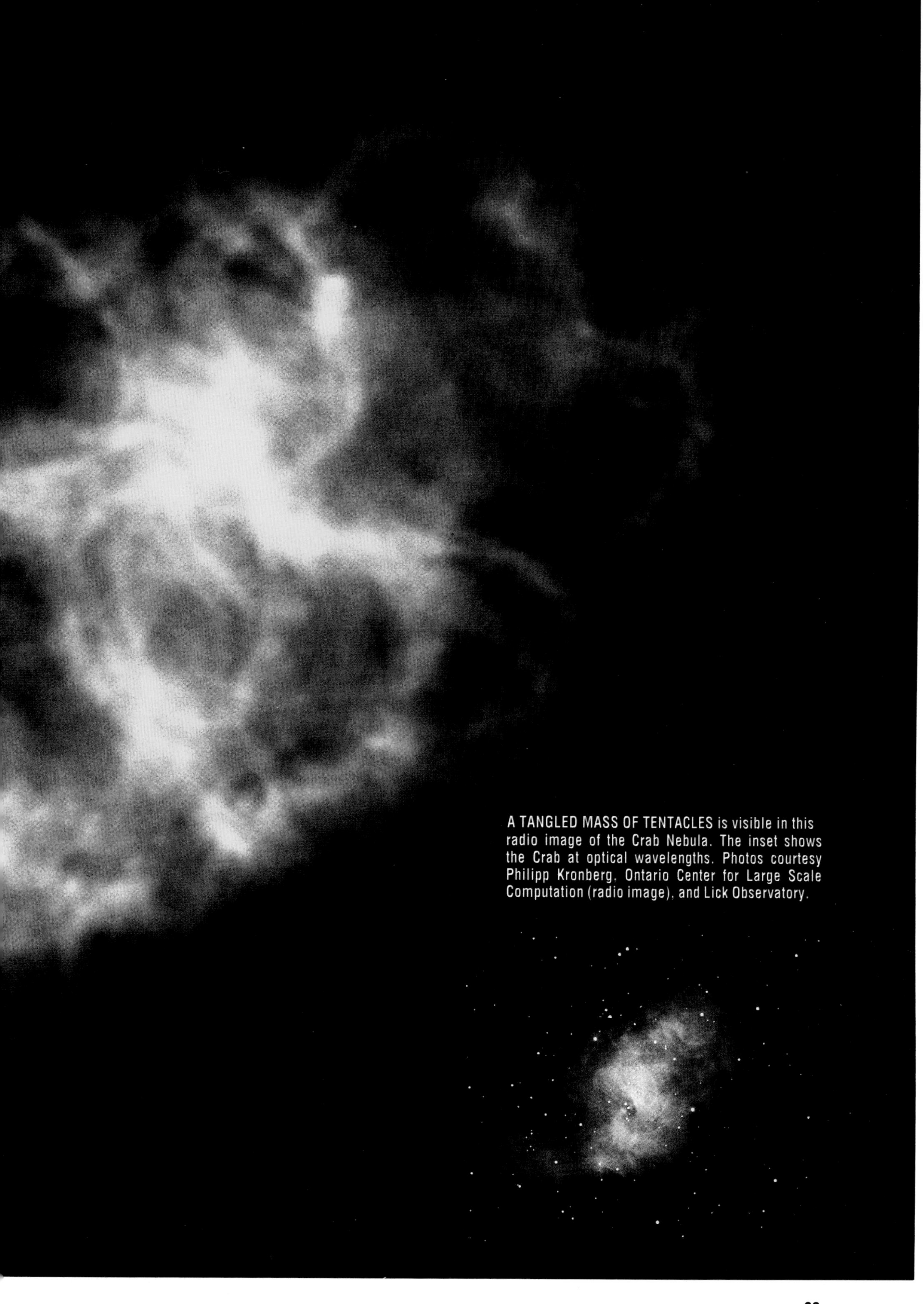

A TANGLED MASS OF TENTACLES is visible in this radio image of the Crab Nebula. The inset shows the Crab at optical wavelengths. Photos courtesy Philipp Kronberg. Ontario Center for Large Scale Computation (radio image), and Lick Observatory.

the object Bevis had originally spotted.

Around 1845, William Parsons, the Third Earl of Rosse, made a sketch of M1 while observing the object with his 72-inch "Leviathon" telescope. His sketch bore a vague resemblance to a giant bug. He christened the object the Crab Nebula.

Astronomers in the early 1900s noticed that the nebula resembled neither Lord Rosse's sketch or a crab. In 1921 Carl O. Lampland of Lowell Observatory compared photographs taken eleven years apart and found that the nebula was expanding at a rate of about 2.25 million miles per hour (1,100 kilometers per second). No wonder the Crab had changed its appearance since Lord Rosse's time.

In the 1950s the Crab Nebula provided the key that led to the solution of the mystery of how cosmic radio sources emit radio waves. Immediately after World War II physicists built the world's first synchrotron at the General Electric Company in Schenectady, New York. Its strong magnetic fields urged electrons into ever more frenzied gyrations around the innards of the device before they were driven headlong into the nucleus of an atom.

When the synchrotron was fired up for the first time physicists watching through a window saw the particle beam begin to glow. At this point in a science fiction movie the heroes would have run headlong for the exits. But the physicists stayed and watched in fascination as the light changed color from dull red at low energies to blue-white when the synchrotron was at full throttle.

Using a Polaroid filter, they found that the light was polarized. This came to be known as synchrotron emission, which is created when electrons traveling near the speed of light (cosmic-ray electrons) encounter magnetic fields. The fields cause the electrons to spiral as they travel. In the process they loose energy, which is radiated away as polarized light. The precise nature and intensity of the emission and the wavelengths over which it can be observed depend on the energy of the electrons and the strength of the magnetic field.

In 1953 the Russian astrophysicist Iosef S. Shklovski suggested that this process was responsible for the light and the radio waves from the Crab Nebula. When the following year the nebula's light was found to be polarized, few doubted that Shklovski was correct.

There remained one annoying fly in the ointment, however. The nebula was emitting so much energy that the cosmic-ray electrons created in the explosion should have stopped creating photons after 200 years. Yet there was the Crab Nebula still shining 900 years after the guest star exploded. How was that possible? In the 1970s the answer was found. A pulsar at the heart of the nebula acts as the engine that provides energy to drive the synchrotron light show. (See "On the Trail of Exotic Pulsars" by Gerrit L. Verschuur, December 1988.)

Images of the Crab Nebula made at radio and visible wavelengths reveal different internal structures and processes at work. The glow of light seen behind and between the filaments in visible light is due to synchrotron radiation. The light from these filaments, however, is generated by a totally different process. The electrons are moving slowly, far below the speed of light, at about 20 kilometers per second. These particles are part of the gas from the original explosion that is kept hot by ultraviolet light generated by the synchrotron process in the more violent parts of the nebula. Thus the visible filaments radiate thermal emission that is more characteristic of the surface of the Sun and objects such as the Orion Nebula.

In a radio image a different set of filaments is revealed. These are magnetic structures that intermingle with the visible ones. The nebula also appears somewhat larger than the optical nebula. Apparently, radio-emitting clouds have expanded faster and reached farther into space than the hot gases.

Bietenholz and Kronberg are especially intrigued by the relationship between the interacting sets of filaments. By observing the detailed structure, motion, and polarization in the Crab, the researchers hope to uncover the dynamical soul of this nebula.

The turbulent filaments of the Crab Nebula fascinate Bietenholz and Kronberg. By comparing two radio images taken five years apart, they can probe the energetics of the Crab supernova explosion. Kronberg admits that the motions they have seen so far, based on the first look at the data, are extremely complex.

Comparing our image of the Crab with one made in 1982 gives us the most detailed look at the physical properties of a supernova remnant ever made.

The Crab Nebula is 6,500 light-years distant and has an angular diameter of 6' by 4'. This translates to about 10 light-years across. The nebula enshrouds from three to six stars. Stars that end their lives as supernovae have to be relatively massive and quite young, something like tens to hundreds of millions of years old. This was not enough time for the emergence of even the simplest life on any planets that might have orbited the star that exploded to produce the Crab Nebula. However, the same is not true of planets around older stars that just happened to find themselves in the vicinity of the explosion. If life ever took hold on any planets orbiting these stars, it has long since ceased to exist. X-rays and ultraviolet light from the explosion, as well as the cosmic-ray electrons in the remnant, are lethal and would have caused total extinction.

This tale of life and death is played out again and again somewhere in the Galaxy every time a supernova erupts. Fortunately for us, there are no suitable stars within hundreds of light-years of Earth that are likely to explode as a supernova any time soon. As regards the fate of life on other planets, it's just a matter of luck as to whether or not it runs into something like the Crab Nebula and life is set back to square one. But once the Crab has faded out of existence and blended with other matter between the stars, life may rise anew, heedless of the fate of countless species that had gone before. □

Gerrit L. Verschuur is a freelance science writer and radio astronomer living in Bowie, Maryland. His last article in ASTRONOMY was "The Peculiar Pulsar in Supernova 1987A," September 1989.

Strange Doings at the Milky Way's Core

New high-resolution radio images of the Galaxy's central arcsecond show swirling clouds of gas, each hundreds or thousands of times bigger than the solar system. Nearby, a red supergiant star is being eroded by a ferocious wind from the Galactic center.

by Robert Burnham

On moonless evenings early this month Sagittarius lies in the southwest. The Milky Way there looks like silver steam rising from the spout of Sagittarius' Teapot asterism, and the stars seem to shine especially bright through the cool evening air. About one palm-width to the right of Gamma Sagittarii, the star at the tip of the Teapot's spout, lies a 5th-magnitude star, X Sagittarii. If you look one finger-width below X Sgr, you will be gazing straight toward the heart of the Milky Way Galaxy. But the Galactic center lies 28,000 light-years away and only radio waves can penetrate the clouds of dust and gas that obscure our view of it.

Three astronomers, Farhad Yusef-Zadeh of Northwestern University, Mark Morris of UCLA, and Ronald Ekers of the Australia Telescope, have used the 27-antenna Very Large Array radio telescope to make highly detailed images of the Galactic center. In particular they focused on the radio source known as Sagittarius A* (pronounced "ay-star"). Sgr A* is not the only radio source close to the center of the Milky Way, but it is the strongest and most compact. Moreover, it appears to dominate large-scale movements, at least locally: gas clouds lying east of it are receding from us, while those to the west are approaching.

These pictures take you from the Sagittarius star clouds all the way in to the plasma clouds flanking Sgr A*. The detail of these images is breathtaking: photo #5 spans just over 1 second of arc. If the solar system were taken to the center of the Galaxy and dropped into the image, its 100 AU diameter (twice Pluto's maximum distance from the Sun) would be as large as the yellow dot.

What, exactly, does lie at the center? Although the actual core object has not been resolved, astronomers

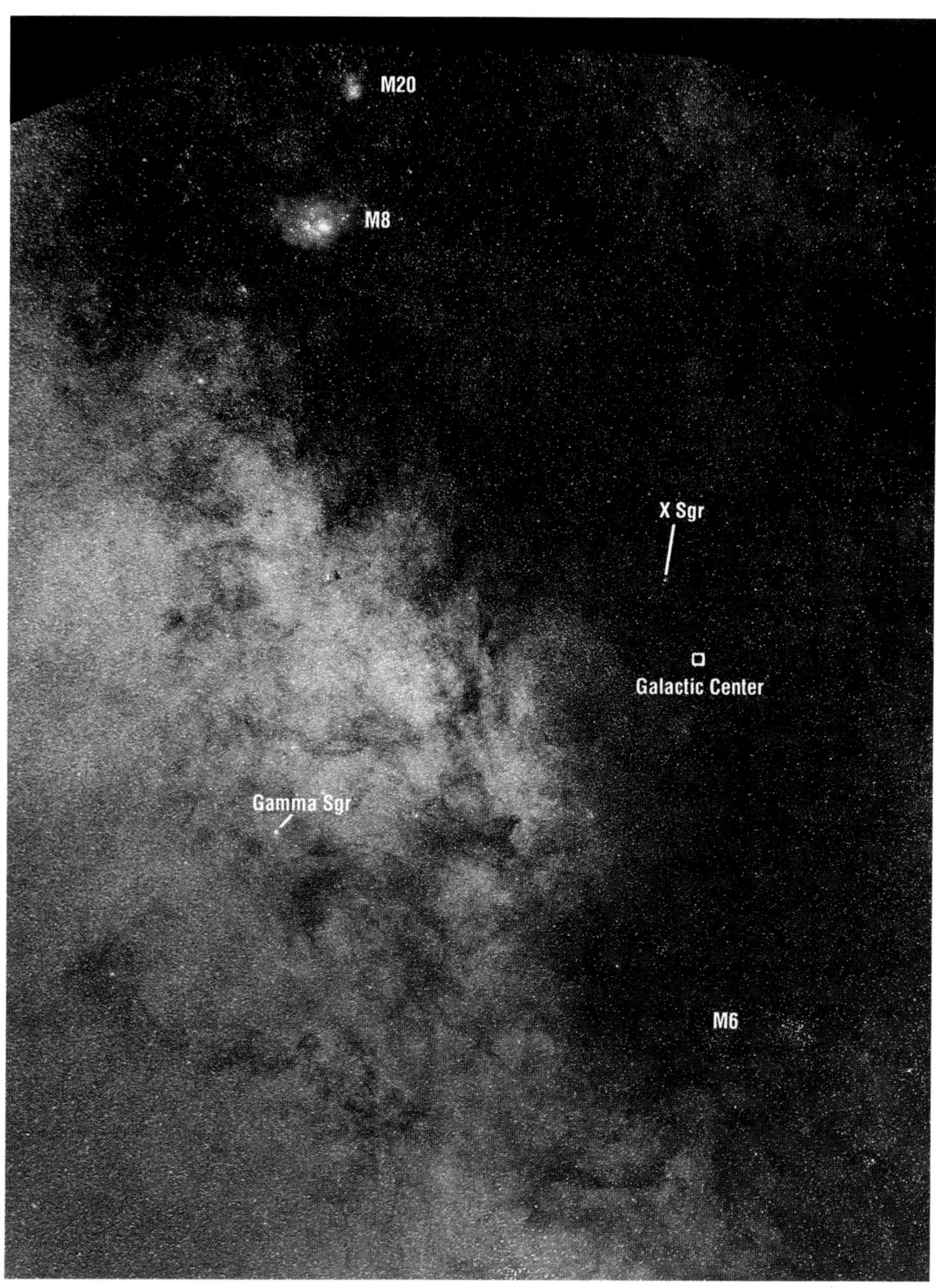

MAGNIFICENT STAR CLOUDS line the Milky Way in Sagittarius, but thousands of light-years deeper into space lies the core of the Galaxy. Photo by Ronald Royer.

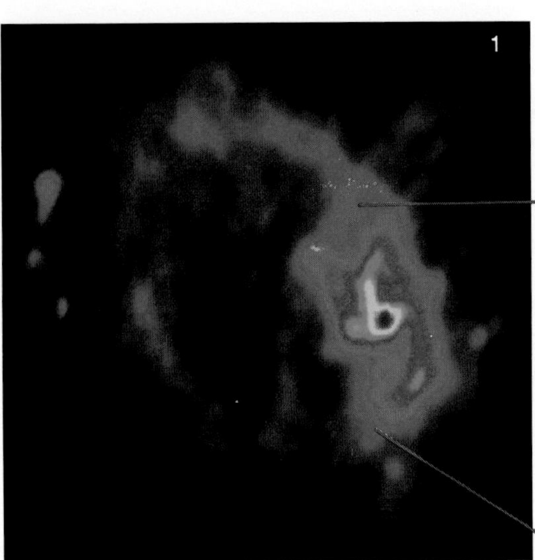

PHOTO #1: Spanning about 5 arcminutes, this giant shell of gas may be a supernova remnant behind the Galactic core, which lies in the "hot spot" at right.

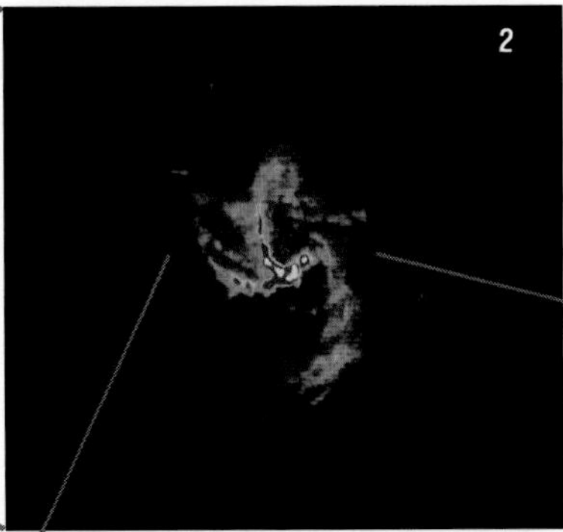

PHOTO #2: The pinwheel structure is about one arcminute across and consists primarily of ionized gas.

PHOTO #3: Enlarged, the pinwheel shows many knots and bright regions. Sagittarius A* marks the source believed to be the Galactic core.

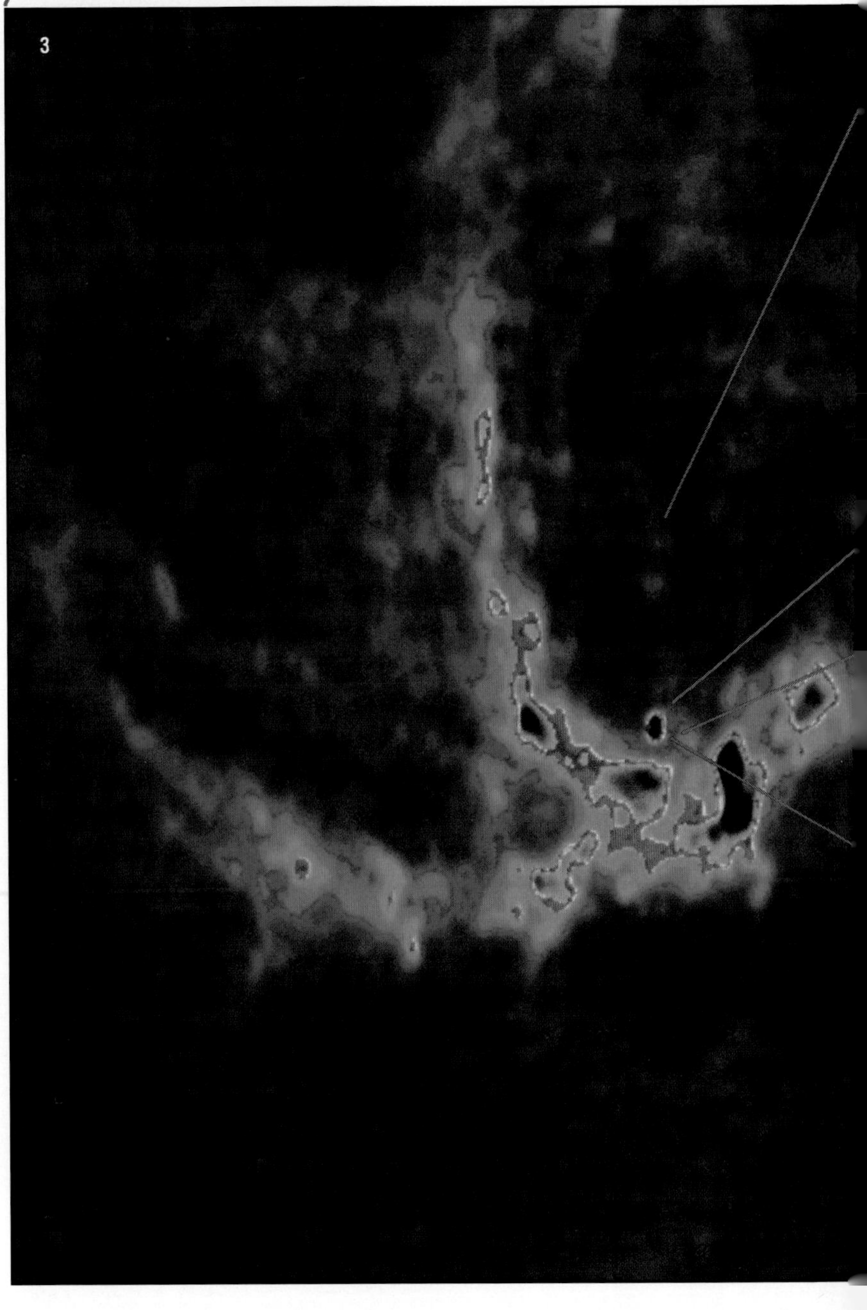

suspect that it is a black hole. Estimates of its mass range from as low as 500 times that of the Sun up to 5 million times. Assuming it exists, the black hole would be surrounded by a sheet of material called an accretion disk, made up of dismembered stars, clouds of gas, and hot dust particles. Material drawn into the black hole would be heated to very high temperatures before it disappeared, thus keeping the accretion disk and vicinity bathed in high-energy radiation. On local scales this radiation ionizes gas clouds, turning them into radio-visible plasma. At a couple of light-years' distance, this "core wind" (plus strong ultraviolet radiation) is eroding a red supergiant star called IRS 7. It is the comet-like object in photo #4.

A tantalyzing question concerns the three-dimensional structure of the features. Which lie in the foreground? Which are distant? The observations here draw a simple two-dimensional map of the features. New observations have measured the radial velocities (speed directly toward us or away) of each object in the photos, but it will take time to process the raw data into answers. An exciting possibility is that the black hole's gravity should be strong enough that any objects lying near-by will be moving quickly. In fact, Yusef-Zadeh, Morris, and Ekers calculate that a similar radio picture taken as soon as 10 years from now would show that the clouds have shifted position.

The core of the Milky Way has long been a mystery, and it retains much mystery still. But the innermost regions of the Galaxy are gradually becoming visible, and accurate maps are finally replacing cloudy speculations. □

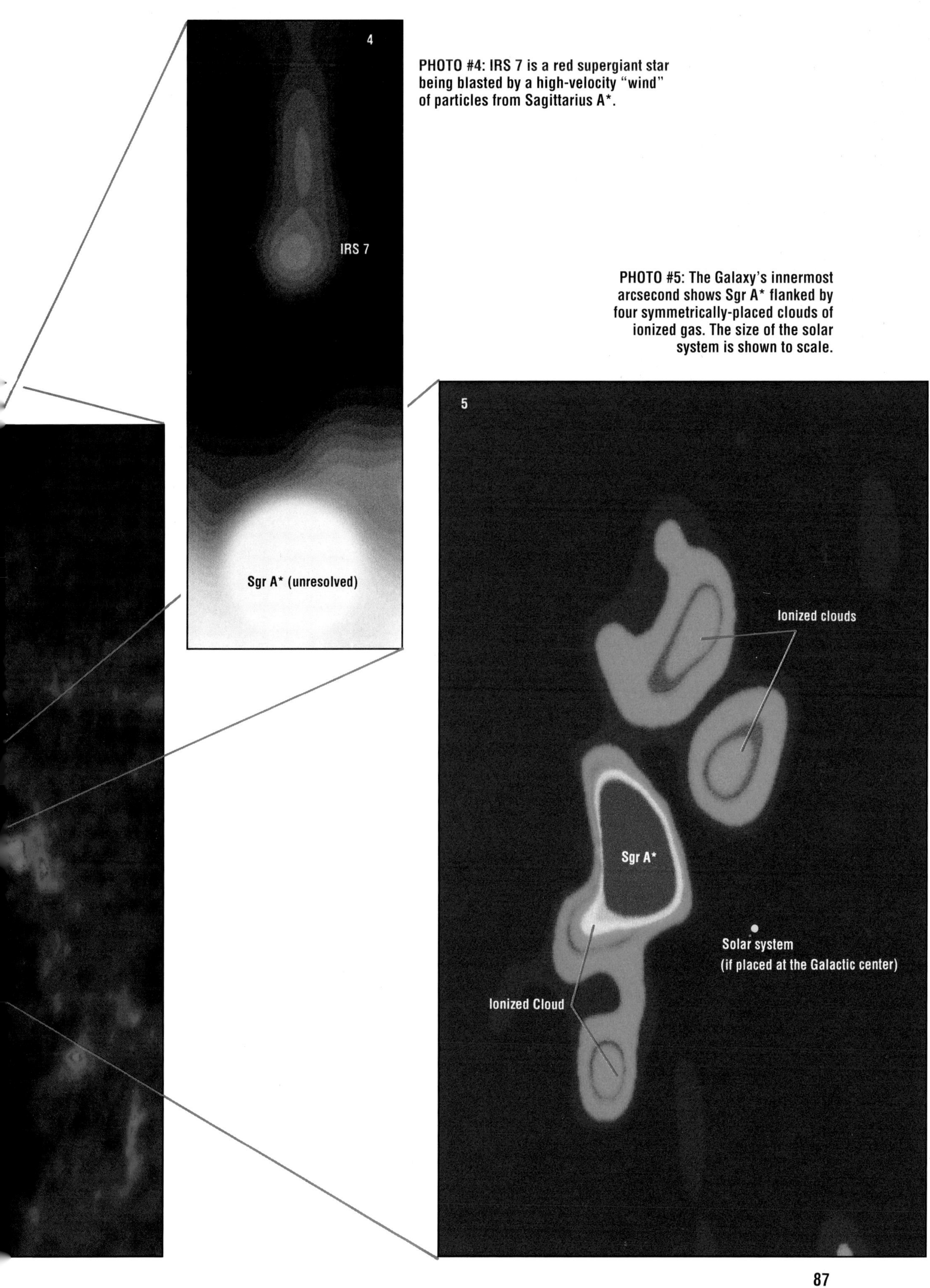

PHOTO #4: IRS 7 is a red supergiant star being blasted by a high-velocity "wind" of particles from Sagittarius A*.

PHOTO #5: The Galaxy's innermost arcsecond shows Sgr A* flanked by four symmetrically-placed clouds of ionized gas. The size of the solar system is shown to scale.

Active Galaxy Centaurus A (NGC 5128)
Photo courtesy National Optical Astronomy Observatories

The Southern Radio Sky Comes to Light

A startling view of the southern Milky Way resulted when a group of astronomers converted an old NASA tracking antenna into a radio telescope.

by Gerrit L. Verschuur

"They took everything not bolted to the floor," said grad student Justin Jonas. "But they *had* to leave the dish."

When NASA abandoned its satellite tracking dish in Hartebeesthoek, South Africa, in the early 1970s, the 26-meter antenna posed both an opportunity and a challenge to the group of astronomers connected with Rhodes University in Grahamstown who inherited the installation. Together with astronomers Eddie Baart and Gerhard de Jager, Jonas and several others outfitted the antenna with sensitive receivers and began to survey the southern radio sky, which was virtually unobserved, especially at short wavelengths.

The photo on pages 40 and 41 shows a false-color view of the Milky Way from the survey, made at 2300 MHz (about 13 cm wavelength). The image extends

ONCE IT TRACKED SATELLITES around Earth, now it probes deep into our Galaxy: this recycled NASA antenna has a new job in the hands of South African astronomers. Below: the Large and Small Magellanic Clouds.

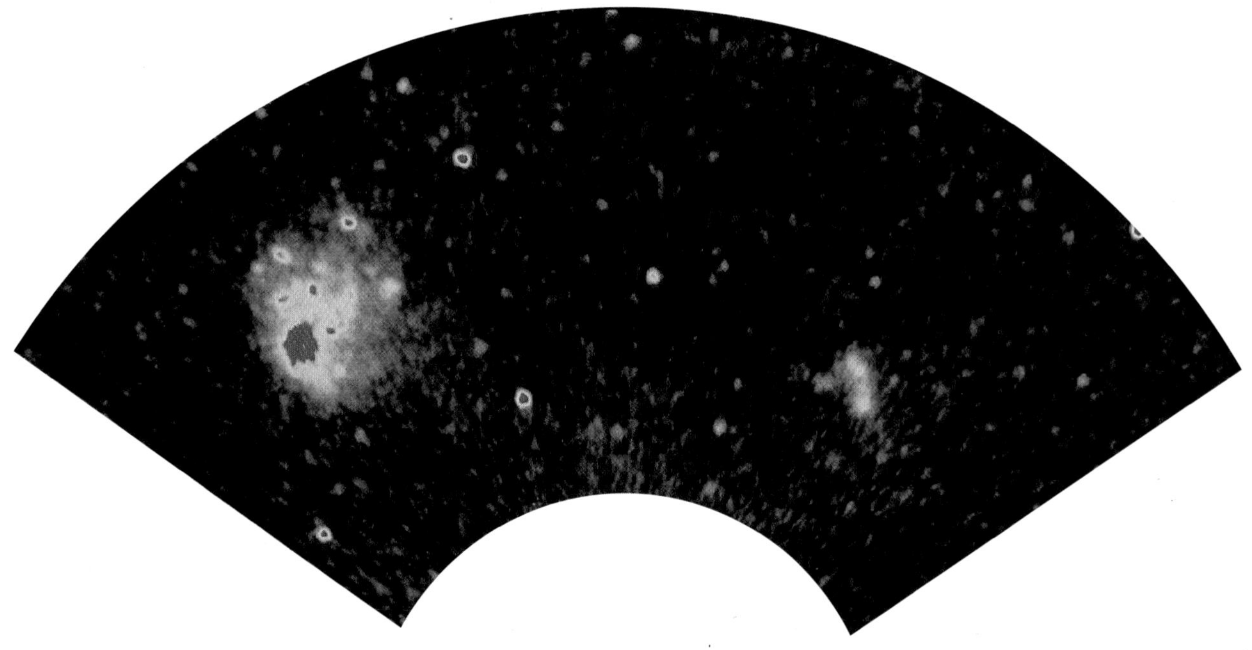

JETS, STREAMERS, AND BUBBLES throng the radio Milky Way just as they do the visible. The Galactic plane forms the thick band across the picture, while supernova remnants and clouds of gas appear projected above it.

from Cygnus (Galactic longitude 70°) on the left through the Galactic center in Sagittarius (0°) to Canis Major (240°) on the right. Vertically, it spans from Galactic latitude 30° south to 60° north.

The radio emission of the Milky Way comes from cosmic ray electrons spiraling around in magnetic fields at nearly the speed of light as they thread a path between the stars. The wide band roughly matches the Milky Way we see with the naked eye. But while the visible Milky Way is produced by stars lying within a couple thousand light-years of the Sun, the radio waves originate 10,000 to 50,000 light-years away.

A great deal of structure shows in the radio survey: filaments (both large and small), knots, and loops. Almost all belong to our Galaxy. One object, however, does not: the small double feature Centaurus A is associated with the enigmatic galaxy NGC 5128, which lies about 15 million light-years away — deep in space beyond the Galaxy's farthest frontier.

This double radio source results from hugely energetic events in the core of NGC 5128. Repeated explosions have hurled twin jets of radio-emitting gas to

The Southern Radio Milky Way (unprocessed)

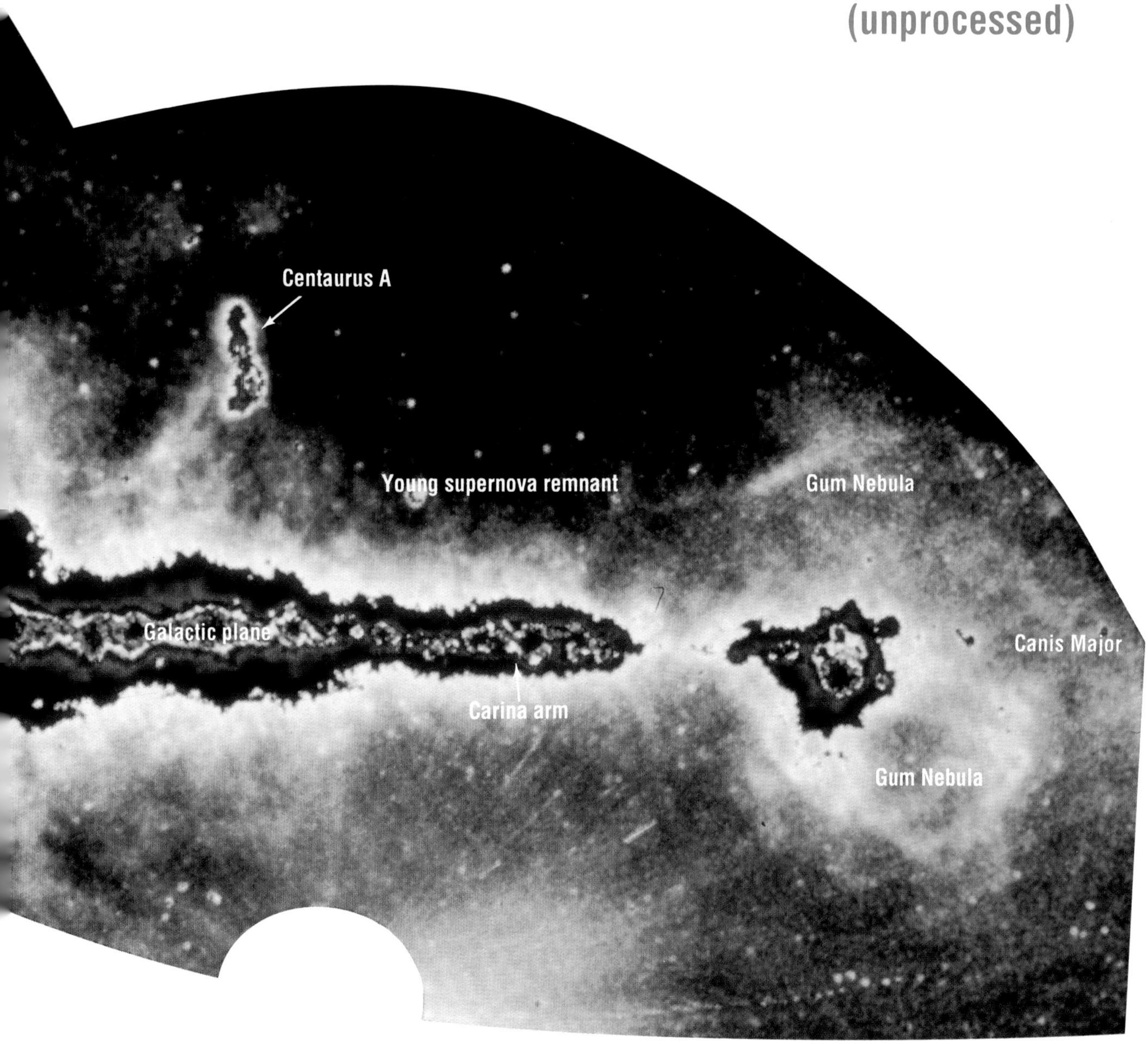

distances of 500,000 light-years. Centaurus A now extends over a distance more than three times the diameter of our own Galaxy. (Astronomers think, however, that the apparent thread linking Centaurus A to the Milky Way is a chance alignment of a nearby feature with the distant galaxy.)

In the Milky Way image a large streamer of emission called the North Polar Spur rises out of the band. Many astronomers believe this is a fragment of energized gas some 330 light-years distant left over from a supernova explosion that occurred about a million years ago. It aligns with the local magnetic field of the Galaxy (see "The Magnetic Milky Way," ASTRONOMY, June 1990).

To the right of the North Polar Spur is a circular structure surrounding a group of hot stars in Ophiuchus, while to its right and a little lower is a loop structure belonging to the Scorpius OB2 stellar association. These features are produced by hot gas surrounding relatively nearby stars. In the center of the image stands a faint and broad ridge of emission called Loop IV, which may be the remnants of another old supernova. Toward the right end of the Milky Way, the Carina spiral arm glows brightly because we are looking down its axis. The small, bright object above and slightly to the left of Carina is a young supernova remnant perhaps a few tens of thousands of years old.

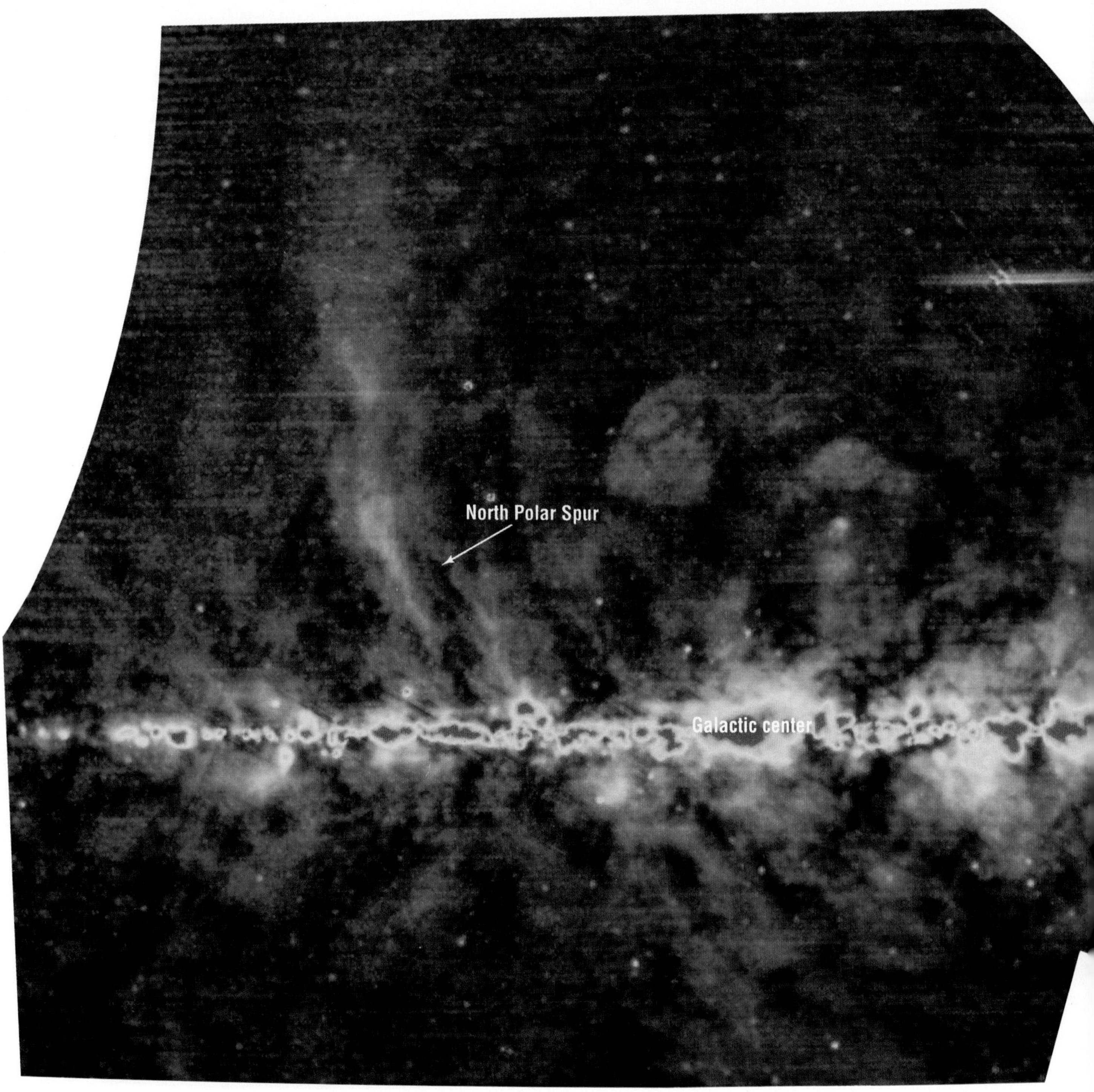

SUBTRACTING AN IDEALIZED GALAXY contour from the image revealed details in the Galactic plane. Note that many of the streamers extending above and below the Galaxy appear to be rooted in knots that lie in the central plane.

The "island" of emission at the right end is the Vela supernova remnant, famed for its pulsar, one of the few visible optically. The Vela SNR lies inside a region of swirling filaments called the Gum Nebula, which surrounds it both above and below the Milky Way. The Gum Nebula (named for its discoverer, Colin Gum) is a big cloud of hot gas heated by young stars, one of which exploded as the Vela supernova.

The image on pages 40-41 conceals most of the underlying structures of the Milky Way. To make them clearer, astronomers subtracted the profile of an idealized galaxy from the observed emission. The result is on these pages. Note how many of the knots in the plane appear to correlate with features extending above and below. Astronomers are now working to understand why these ridges, filaments, and loops occur where they do. When the survey is complete — it has been underway for more than 11 years — the Rhodes University astronomers plan to repeat their observations at 6 cm wavelength. This new look at shorter wavelengths will reveal finer details.

Over the last decade Rhodes University astrono-

The Southern Radio Milky Way (detail enhanced)

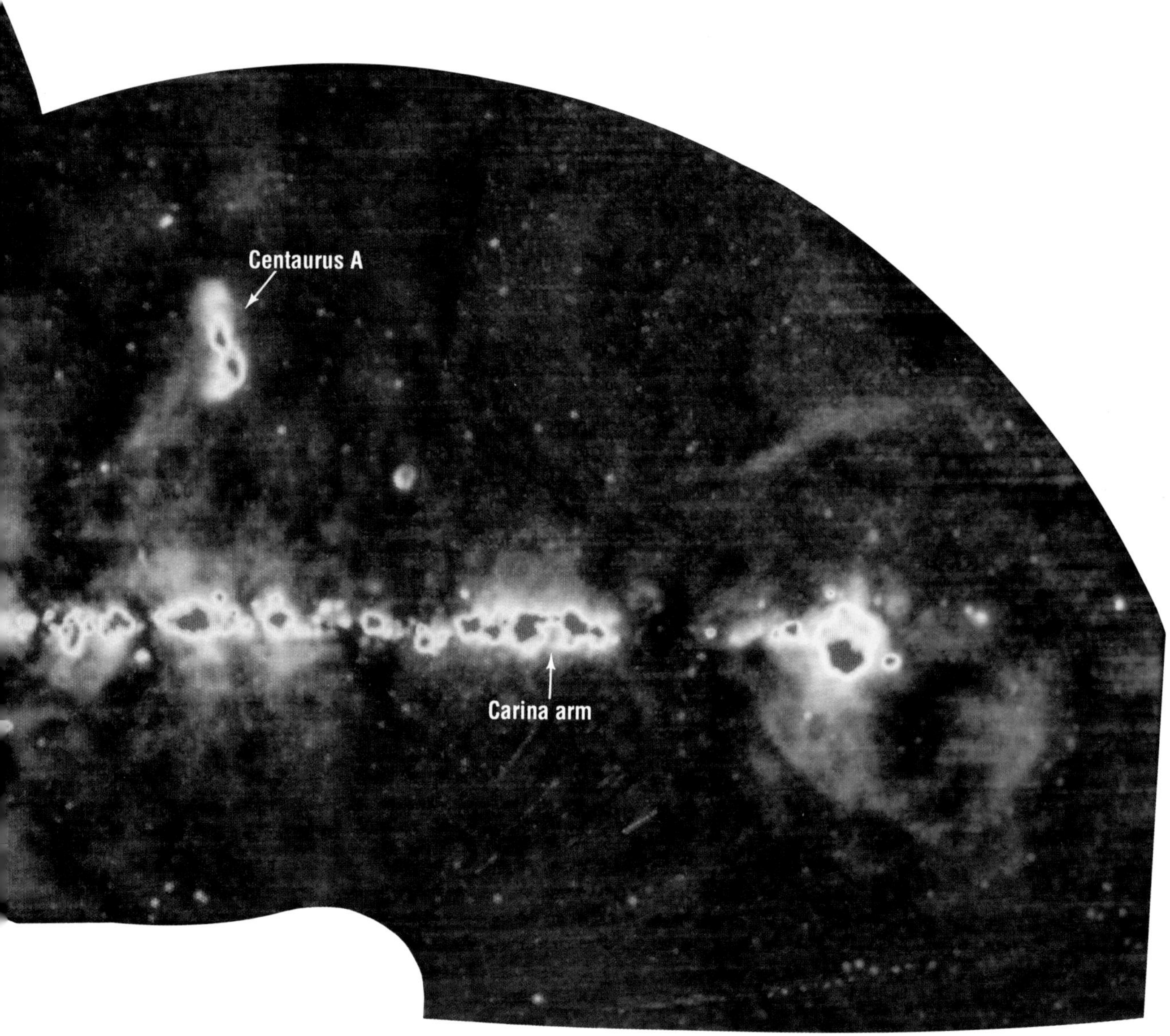

mers have steadily nursed sections of their huge database of observations through the computers with the aid of graduate students including Mike Wright, Justin Jonas, and currently Beate Woermann. The telescope's resolution of 20 minutes of arc means that vast numbers of observations are required to complete such a survey. The difficulty is compounded because most observations must be repeated to eliminate "noise" from manmade interference: significant — and growing — amounts of data are being lost due to satellite transmissions, which swamp the natural signals the way light from an errant flashlight can ruin an astrophoto.

Very few radio scopes have probed the southern Milky Way and the great expanses around the south celestial pole, where the Magellanic Clouds reign supreme. When an out-of-work radio dish met a team of determined astronomers, the result was a premier scientific instrument for exploring these skies. It is also an example that other nations, unable to afford building their own large radio telescope dishes, may hopefully follow. □

Gerrit L. Verschuur is a freelance science writer and researcher in Bowie, Maryland. He is the author of Interstellar Matters *(Springer-Verlag) and "The Magnetic Milky Way" in the June 1990 issue.*

A Journey with Light

Come on a 12-billion-year journey as the light from a distant quasar travels to Earth.
by Govert Schilling

Springtime in the Canary Islands. During the day exotic birds sing from the palms as the turquoise water washes up on the beach. Since 1986 telescopes high atop the islands' Roque de las Muchachos mountains have collected light from distant celestial objects. Tonight an astronomer is using the largest telescope at the observatory, the Anglo-Dutch William Herschel Telescope, to photograph a distant quasar. Once the spectrograph has been placed in the focal plane of the telescope, she aims the telescope by computer. The light from the quasar passes through the telescope and is recorded by an electronic detector.

For the astronomer this is a routine observing run. But what about the light from the quasar? What of its experiences before being captured by the electronic detector on a warm spring night on Earth?

The feeble light our astronomer records left the quasar 12 billion years ago. The particles of light that arrived on Earth tonight began their journey at a time when Earth did not yet exist.

Traveling at 186,000 miles per second, the light from the quasar carried a record of the conditions in a very distant part of the universe. Quasars, discovered nearly thirty years ago as starlike objects with peculiar spectra, appear to be the most distant objects known. Although they are not well understood, quasars are thought to be the highly energetic cores of young galaxies, possibly containing supermassive black holes. In any event, the brightness of quasars at extreme distances means they are tremendously energetic objects. Twelve billion years ago a wave of light from our quasar set out on a journey through space to an unknown destination. After the light left the quasar, many years passed — tens, hundreds, and thousands. The light lost none of its speed. Each second it moved fast enough to circle Earth's equator seven and a half times. But even after ten thousand years of travel, the light still hadn't left the quasar's host galaxy. Only when it emerged from this galaxy did the light's journey between the galaxies begin.

One, two, and three billion years slowly passed. Our own galaxy, the Milky Way, was forming into a spiral-shaped disk in which billions of stars softly turned on. Throughout these billions of years the light from the quasar sped through the emptiness of space. The light passed across vast regions of the universe called supervoids. These areas contain nothing but black, cold space. Then the light passed through clus-

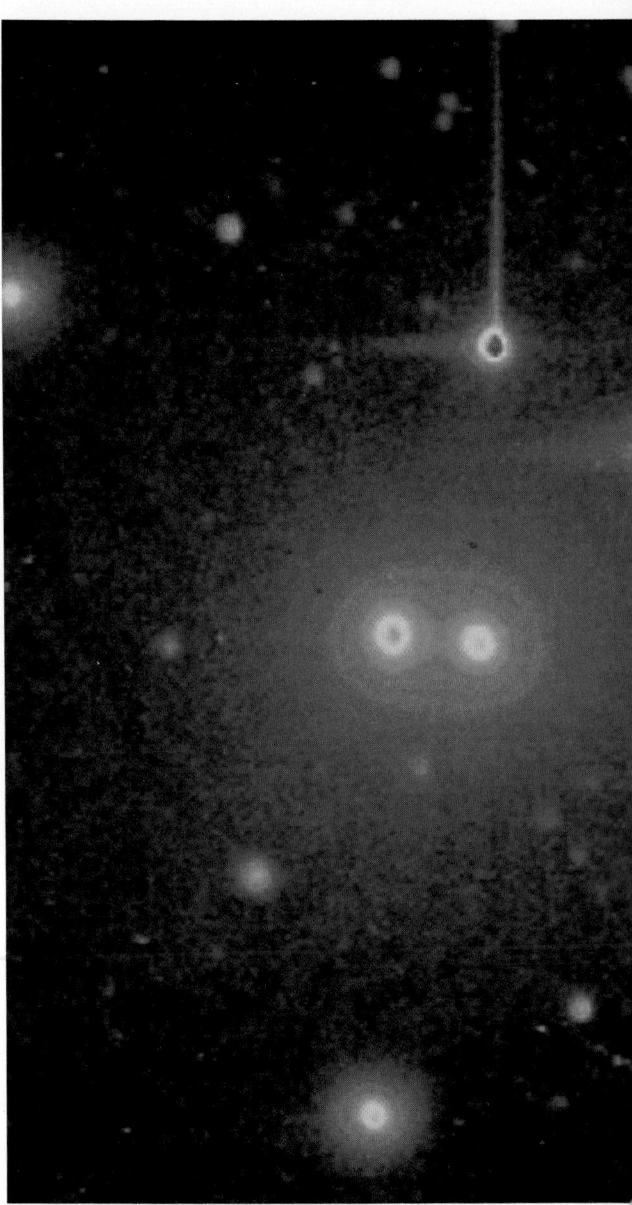

A LIGHT-SPEED JOURNEY through vast corridors of space brought the light of a quasar to Earth. From the quasar's point of view reaching Earth was the culmination of an extraordinary flight. NRAO photo.

Bruce Bond

NOAO

A TWELVE-BILLION-YEAR TREK brought the wave of quasar light from the edge of the universe to Earth. As the light passed through the Coma Cluster of galaxies (above) 350 million years ago, life on Earth was diversifying. Fish were numerous and tree-ferns inhabited the shores. As the light passed through nearby galaxies like NGC 4501 (right), dinosaurs encountered rough times in the wake of a dust storm produced by the collision of an asteroid with Earth. Soon thereafter, *Homo habilis* inhabited the trees. Some 10,000 years ago the quasar's light encountered numerous Milky Way objects like IC 2220, the Toby Jug Nebula (opposite page). Early humans gave birth to astronomy as they gazed toward the nighttime sky and wondered about the stars.

Rudolf Schild, Harvard-Smithsonian Center for Astrophysics

Anglo-Australian Telescope Board

Twelve billion years ago a wave of light from a distant quasar set out on a journey through space to an unknown destination.

GIANT TELESCOPES like the William Herschel Telescope in the Canary Islands trap ancient light that lets us peer deep back in time toward the very beginning of the universe. Royal Greenwich Observatory photo.

ters or superclusters of galaxies. Every few million years it briefly visited a galaxy. After 6 billion years the quasar's light had covered half the distance to Earth. But in what is now the solar system, there was not yet any trace of Earth or even the Sun.

Seven billion years after the quasar's light was emitted, a cloud of interstellar gas and dust slowly contracted and consolidated, and the Sun and planets formed. During the time Earth formed, the quasar's light passed through a supercluster of galaxies. Five hundred million years later, as the light sped through the darkness of intergalactic space, Earth's first single-celled organisms developed. The evolution of an intelligent and inquisitive mammal had begun.

As life on Earth evolved into multicelled organisms, the light from the quasar relentlessly rushed onward by 6 trillion miles each year. By now the quasar's light had entered our region of the universe. Three hundred fifty million years ago it was about as far away as the Coma Cluster of galaxies. Fifty million years later the quasar's light was closer than the Perseus Cluster of galaxies. On Earth the first amphibians emerged. Numerous fishes inhabited the oceans, and ferns and bracken covered the continents.

To a twentieth-century observer Earth would have been unrecognizable. The present distribution of land and sea bears no resemblance to that of a few hundred million years ago. As the light from the distant quasar approached the Virgo Cluster of galaxies, the primeval continent of Pangaea began to slowly break apart and Earth's crustal plates gradually drifted apart. By now, the quasar's light had journeyed 11.8 billion years — 98.5 percent of the distance to Earth. The journey was nearing its end. Even so, on Earth there was still no sign of mammals. As the continents began to move apart the first dinosaurs were evolving, to be followed only 50 million years later by the first mammals.

Dinosaurs held absolute sway over our planet for tens of millions of years. But their rule came to a sudden end about 65 million years ago when an asteroid or comet hit Earth. The atmosphere was laden with thick clouds of dust for many years. Sunlight was blocked, food chains were broken, and countless species died out. The cosmic catastrophe thus cleared the way for the evolution of mammals. By then, the quasar's light had covered 99.5 percent of the distance to Earth and was in the Virgo Cluster.

The onward-speeding light reached the Local

We had received a message from an incredibly powerful and distant place that we will never see exactly the same way again.

Group of galaxies and was soon closer to Earth than the Andromeda Galaxy. On Earth the family of mammals was large and rich in species, from the huge mammoths to the relatively small *Homo habilis* that lived in the trees of extensive tropical forests.

Homo erectus appeared as the quasar's light passed the Magellanic Clouds, two small galaxies that revolve about the Milky Way. Later, subtle periodic variations in Earth's orbit caused ice ages that forced *Homo sapiens* to lead a nomadic existence. The light passed the Milky Way's center, leaving a mere 25,000 light-years between the quasar's light and Earth. The light passed stars, star groups, and gas clouds. By the time the light was 10,000 light-years from Earth, the last ice age had come to an end. The quasar's light had covered 99.9999 percent of the distance it would travel. By now there may have been human beings looking up at the night sky, thinking about the thousands of points of light they saw. Astronomy was born.

Civilizations rose and fell. Wars were won and lost yet the problems that created them persisted. People discovered writing and the first records of astronomical observations were made. The quasar's light approached unrelentingly. Greek civilization flowered and the first steps were taken on the path to modern science. At about 500 A.D. the quasar's light passed the Orion Nebula. Eleven hundred years later the light was as close as the Pleiades star cluster, a little less than 400 light-years from Earth. Galileo Galilei became the first person to turn a telescope toward the sky and humanity's discovery of the universe was underway. It was January 1610.

The development of the telescope proceeded alongside astronomy itself. Astronomers such as William Herschel, Charles Messier, and Clyde Tombaugh discovered planets, nebulae, and star clusters. Telescopes got bigger and bigger and mirrors were used to focus starlight rather than just lenses. By 1948 the light from the quasar had passed one of the brightest stars in the sky, Capella. That year the Hale Telescope's 5-meter mirror began probing the skies from Palomar Mountain, California. The moment was fast approaching!

By 1963 the quasar's light was closer than the bright, blue-white star Vega. This was the year astronomer Maarten Schmidt deduced that some unusual radio sources that looked like stars were actually incredibly distant. They came to be called quasistellar objects, or simply quasars. No one understood how they generated such great amounts of energy. Quasars became the focus of intense research.

Meanwhile, the light from our quasar sped closer. During the summer of 1985, as the Herschel Telescope was being finished, the quasar's light put the last of the stars, Alpha Centauri, behind it. The following year the Herschel Telescope was shipped from Britain to La Palma. Engineers spent months setting up and testing the instrument and each day the light from the distant quasar closed in by more than 16 billion miles. In 1987, the Herschel Telescope saw first light.

Today dawned and the Dutch astronomer prepared to observe the quasar. By noon she had checked all the details of the observing run. At midafternoon the quasar's light passed a small unmanned spacecraft traveling beyond the orbit of Pluto. It was *Pioneer 10,* a probe launched from Earth in 1972. *Pioneer 10* had spent eighteen years moving out of the solar system at a speed of 16 miles per second. But the quasar's light would cover the distance of *Pioneer 10's* journey in just a few hours.

By early evening the quasar's light had crossed Pluto's orbit. At 9:15 p.m. the light crossed the orbit of Uranus, and at 10:30 it traversed Saturn's orbit. Forty-five minutes later the light passed inside Jupiter's orbit. At that moment the astronomer placed the spectrograph into the telescope's focal plane. She then left the dome to take a five-minute coffee break, during which time the light from the quasar hurtled some 60 million miles closer. The quasar's light shot past the Moon and spilled onto the floor of the dome. The astronomer moved the telescope and the light entered the telescope. After twelve billion years the quasar's light struck the astronomer's light-sensitive detector.

Another observation in an endless routine of data collecting, the astronomer thought. But in reality it was the end of a momentous journey, a trip through clusters and supervoids, past galaxies, nebulae, and solitary stars. We had received a message from an incredibly powerful and distant place that we will never see exactly the same way again. □

This article is based on material from Dimensies in de Natuur by Wilfred Kruit and Govert Schilling (Aramith Publishers, Amsterdam, 1987). It was translated from Dutch by Rosalind Melis.

DARK MATTER PERMEATES the universe, from clusters of galaxies to the galaxies themselves. Even the destiny of the universe may be controlled by this unseen matter.
Painting by Michael Carroll.

COSMOLOGY

Shedding Light on Dark Matter

We can't see it, but we know it's all around us. It's called dark matter.

by Richard Monda

Fritz Zwicky uncovered a mystery in 1933, one astronomers are still trying to solve. Zwicky, a Caltech astronomer, found that the universe contains more matter than we see. This unseen matter deeply interests astronomers because its gravitational influence may control the destiny of the universe.

The first hint that dark matter exists came from observations of galaxy clusters. Zwicky measured the motions of galaxies in the Coma cluster and realized that the individual galaxies were moving too quickly for the cluster to stay together for a long period of time. The motion of each member of the cluster should have caused the cluster to fly apart long ago. Yet even backyard telescopes show that the cluster remains intact.

Zwicky concluded that the cluster must be ten times more massive than it appears in order for the cluster to remain gravitationally bound. This enormous mass discrepancy indicates that ninety percent of a cluster is invisible. Also, because matter in the universe clumps to form galaxies and galaxy clusters, the mass discrepancy implies that 90 percent of the *universe* is invisible.

At first, astronomers called the material "missing mass," but the term is misleading. Zwicky's observations show that the mass is there. "Dark matter" is a better term because it is the light emitted by the material that is missing.

There are cosmological implications to the existence of dark matter. The standard cosmological model predicts that the universe continues to expand and cool after the initial big bang. If, however, there is sufficient mass in the universe, the expansion will be slowed and possibly even reversed. Therefore, the amount of dark matter controls the expansion — and destiny — of the universe.

In spite of the obvious importance of dark matter, astronomers do not know the form of this matter. The list of discarded candidates gets longer each year, yet the form of dark matter remains murky. But we know where it is. It permeates the space between galaxies and the galaxies themselves.

Dark Matter in Galaxies

Just as the motions of galaxies in clusters indicate that matter lurks unseen in the cluster, the motions of stars give evidence for dark matter in galaxies. Vera Rubin, an astronomer at the Carnegie Institution of Washington, measures how quickly stars rotate about the center of galaxies. Plotting the stars' rotational speed versus their distance from the galactic center yields a rotation curve. Our Galaxy's rotation curve reveals that most of the galaxy — perhaps out to 300,000 light-years — rotates like a rigid wheel. Rubin's work conducted over the last two decades shows that other galaxies behave the same way.

Her results were surprising because astronomers expected stars to orbit a galaxy in the same way that planets orbit the Sun. Mercury, the innermost planet, speeds along at 30 miles per second, while the outermost planet, Pluto, plods at 3 miles per second. In contrast, stars in the Milky Way Galaxy move at roughly 150 miles per second, regardless of their distance from the center. The only possible explanation is that the visible part of the Galaxy is surrounded by large amounts of unseen matter.

Besides rotation curves, there is other evidence for dark matter. Astronomers find from the orbital motions of globular clusters that the Galaxy's mass is much larger than that of visible matter. The conclusion is inescapable. Ninety percent of the mass of the Galaxy must be in the form of dark matter. The most likely location for this mass is in the halo surrounding the Galaxy.

The ratio of a galaxy's mass to its luminosity, or energy output, also supports the existence of dark matter. For stars like the Sun, the ratio is roughly one (one solar mass divided by one solar luminosity). Hot stars have ratios less than one because they emit lots of energy for their mass. Conversely, the ratio is greater than one for low-mass, faint red dwarf stars.

Visible matter in our Galaxy, which largely contains red dwarfs, has a mass-to-luminosity ratio of 2. But the ratio of total mass to luminosity is five times higher. The high total mass-to-luminosity ratio is found also in other galaxies and ranges from 10 to 30. Binary galaxy systems and clusters of galaxies have ratios as high as 300. Large amounts of dark matter are needed to explain these ratios.

After 60 years of observation, in short, astronomers conclude 1) that galaxies and galaxy clusters must contain 10 times more matter than we can see and 2) that mass is most likely contained in the halo surrounding galaxies and in the spaces between galaxies. What isn't known, though, is the form of this invisible matter.

What Dark Matter Isn't

There are so many possible forms for dark matter that it is easier to state what it *isn't*.

We know it isn't atomic hydrogen gas because we would detect

A HIGH MASS-TO-LIGHT RATIO means most of the galaxy's mass is dark matter. M101 photo by Tony Hallas and Daphne Mount.

GALAXIES ROTATE at nearly constant speed rather than like the planets in our solar system.

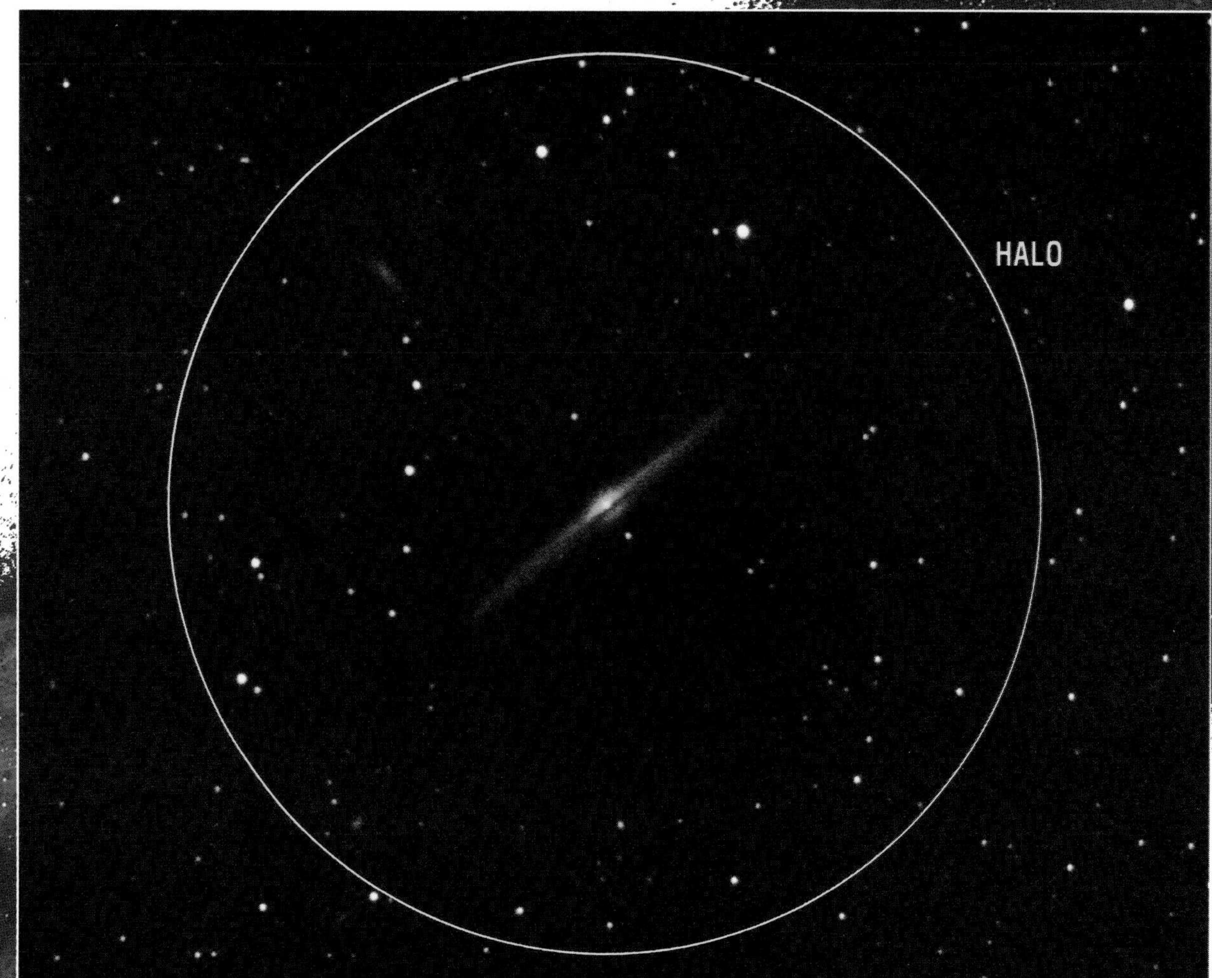

MASSIVE, FAINT HALOS SURROUND galaxies with dark matter. NGC 4565 photo by Jim Baumgardt.

radio emission from gas clouds. Likewise, optical radiation from ionized hydrogen gas clouds would betray their presence. Even molecular hydrogen would show absorption lines in the spectra of distant objects as light from these objects passes through the gas cloud.

We know it isn't dust, for large amounts of dust would obscure light from distant galaxies more than is observed. Dark matter can't be stellar mass black holes or neutron stars either. Gas falling into these objects would emit more X rays than satellites have observed.

None of these radiation signals are detected by astronomers, so dark matter is not in the form of gas, dust, stars, or black holes. However, there are many other possible candidates such as brown dwarfs, black dwarfs, Jupiter-sized bodies, or exotic particles.

Brown and Black Dwarfs

Most stars visible in the Galaxy are faint, low-mass red dwarfs. But astronomers predict that even smaller objects, brown dwarfs, exist that are fainter and cooler than red dwarf stars. Brown dwarfs are at least 80 times more massive than Jupiter. They are more than just large planets, however. They form as stars do from large gas clouds and produce energy for a short time through nuclear reactions.

To explain dark matter in the Galaxy by brown dwarfs alone, there would have to be at least 200 times the estimated number of red dwarfs. Far fewer probably exist, since the efforts of several groups of astronomers have produced only a handful of brown dwarf candidates (see "Do Brown Dwarfs Really Exist," April 1989). Some astronomers conclude that while brown dwarfs may contribute to dark matter, they do not exist in sufficient quantity to provide the large amount of observed dark matter.

Black dwarfs are another possibility. These objects are not failed stars like the brown dwarfs but stars that have run through their entire life cycle. A black dwarf is the ultimate endpoint for a low-mass star like the Sun. Such a star is black because it no longer emits light and, hence, is impossible to detect directly. But that isn't the biggest problem with black dwarfs: it takes too long to make one.

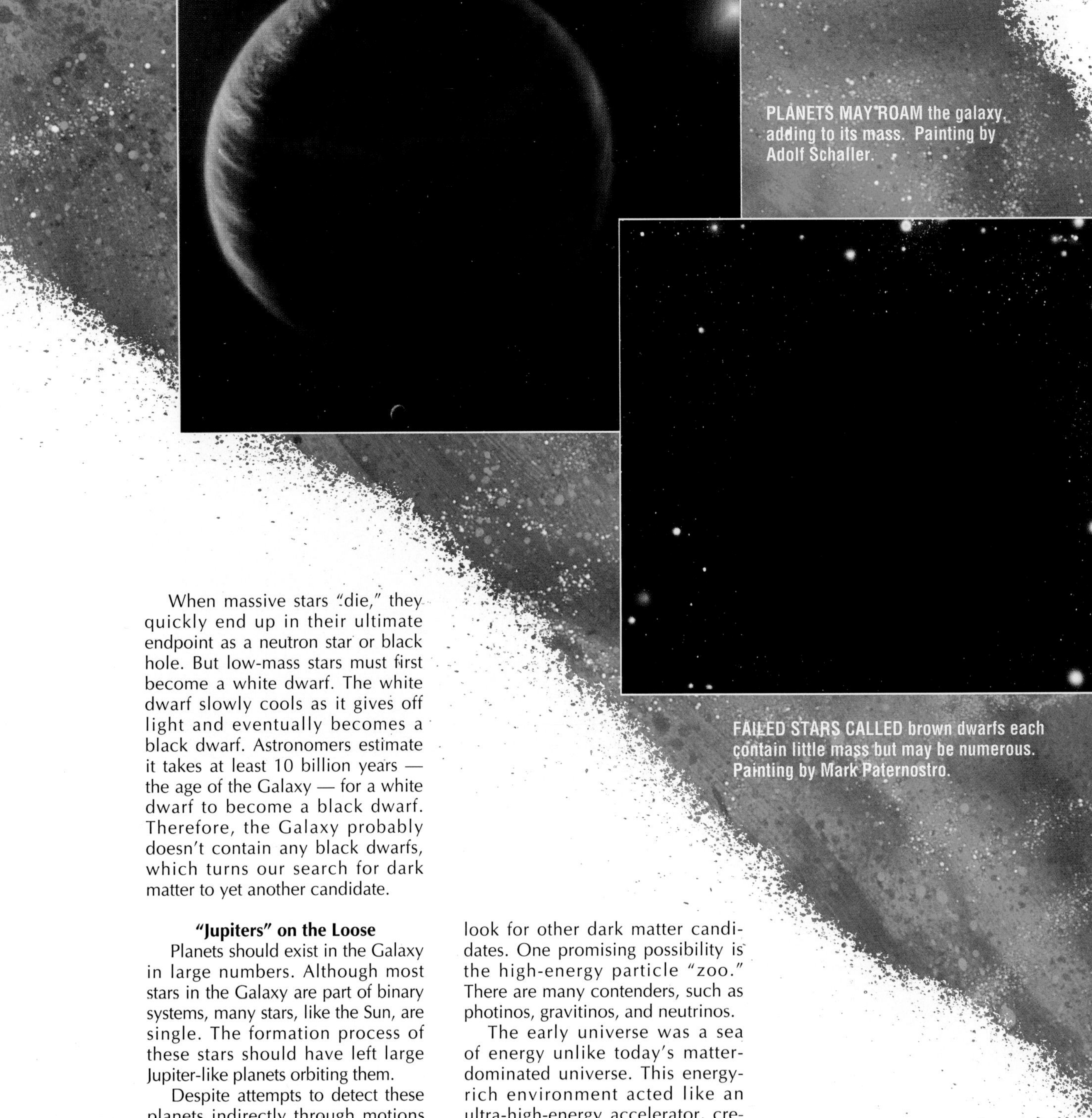

PLANETS MAY ROAM the galaxy, adding to its mass. Painting by Adolf Schaller.

FAILED STARS CALLED brown dwarfs each contain little mass but may be numerous. Painting by Mark Paternostro.

When massive stars "die," they quickly end up in their ultimate endpoint as a neutron star or black hole. But low-mass stars must first become a white dwarf. The white dwarf slowly cools as it gives off light and eventually becomes a black dwarf. Astronomers estimate it takes at least 10 billion years — the age of the Galaxy — for a white dwarf to become a black dwarf. Therefore, the Galaxy probably doesn't contain any black dwarfs, which turns our search for dark matter to yet another candidate.

"Jupiters" on the Loose

Planets should exist in the Galaxy in large numbers. Although most stars in the Galaxy are part of binary systems, many stars, like the Sun, are single. The formation process of these stars should have left large Jupiter-like planets orbiting them.

Despite attempts to detect these planets indirectly through motions of the star introduced by the planet's gravity, there are presently no confirmed detections of planets. (Direct sightings will be attempted with the Hubble Space Telescope when the telescope optics are corrected.) This suggests that either there are fewer planetary systems than first thought or the planets have low mass. In either case, planets alone are not the solution to the dark matter problem.

Exotic Particles

The absence of brown dwarfs, black dwarfs, and other planetary systems has driven astronomers to look for other dark matter candidates. One promising possibility is the high-energy particle "zoo." There are many contenders, such as photinos, gravitinos, and neutrinos.

The early universe was a sea of energy unlike today's matter-dominated universe. This energy-rich environment acted like an ultra-high-energy accelerator, creating nuclear particles we presently only envision. The high-energy phase of the universe's history ended less than a second after the big bang, but some of these exotic particles may still exist in the universe. Of particular interest is the class of particles called WIMPs. These Weakly Interacting Massive Particles are difficult to detect because they do not easily interact with other forms of matter. This low interaction makes them ideal candidates for dark matter.

Like other candidates for dark matter, most of these predicted particles have not been detected as yet. One exception is the neutrino. The Sun produces countless billions of neutrinos in its core as hydrogen is converted into helium. However, astronomers detect fewer neutrinos than predicted by models of the Sun's interior. One possible explanation is that the neutrino has mass — like protons and electrons but unlike the massless photons — and, therefore, interacts more strongly with other material inside the Sun than theory predicts. Currently, though, researchers believe that even if the neutrino has mass, it is too small to contribute significantly to dark matter.

Astronomers and physicists have joined in the search for other predicted particles. But in contrast to other candidates, such as brown dwarfs, not having detected these particles keeps interest high. Their large mass, potentially high abundance in the universe, and lack of interaction with visible matter (which keeps them in the dark) make these particles prime candidates for dark matter.

Cosmology and Dark Matter

Despite the many possibilities, astronomers still do not know the form of dark matter. But even without knowing its form, the cosmological implication of dark matter is clear. If there is sufficient material, the mutual gravitational attraction of objects in the universe will eventually halt the present expansion and cause the universe to collapse. On the other hand, if there is not enough material, gravity is too weak to slow the present expansion and the universe will expand forever. If the amount of mass is just right, gravity will bring the expansion to a halt but will be too weak to make the universe collapse. The expansion rate in this last case is called the critical speed.

The visible mass contained in the universe is 1,000 times less than needed to halt the expansion and even including present estimates of dark matter, the mass is still 5 to 10 times too small to cause the universe to collapse. To cause the collapse, the mass-to-luminosity ratio of galaxy clusters must be at least seven times greater than the largest ratio presently observed. Yet inflationary models of the universe predict that the expansion rate should be nearly the critical speed. Perhaps there is even more dark matter yet to be found in the universe.

How pervasive is dark matter? It is directly detected in galaxies and galaxy clusters. It might also exist in the space between clusters of galaxies and in the space between clusters of clusters. Astronomers have detected voids in the larger fabric of the universe where galaxies are not found. If dark matter exists in these regions, a major problem in astronomy may shift from how matter clumps in otherwise uniform space to what causes galaxies to form only in some regions of space.

As crucial as unseen matter is to our understanding of the size, shape, and mass of the galaxies and to the structure and destiny of the universe, we know only where it is located. There doesn't seem to be enough of it to stop the expansion of the universe and we know almost nothing about its form. Only one thing is really clear at this time. Astronomers are just beginning to shed light on dark matter. □

Richard Monda is the director of the Schenectady Museum Planetarium in New York.

EXOTIC PARTICLES LEFT OVER from the Big Bang may be the source of the dark matter.

ASTRONOMERS PRESENTLY BELIEVE there isn't enough mass in the universe, even with dark matter, to stop its expansion. Painting by Victor Costanzo.

The Case for Density Waves

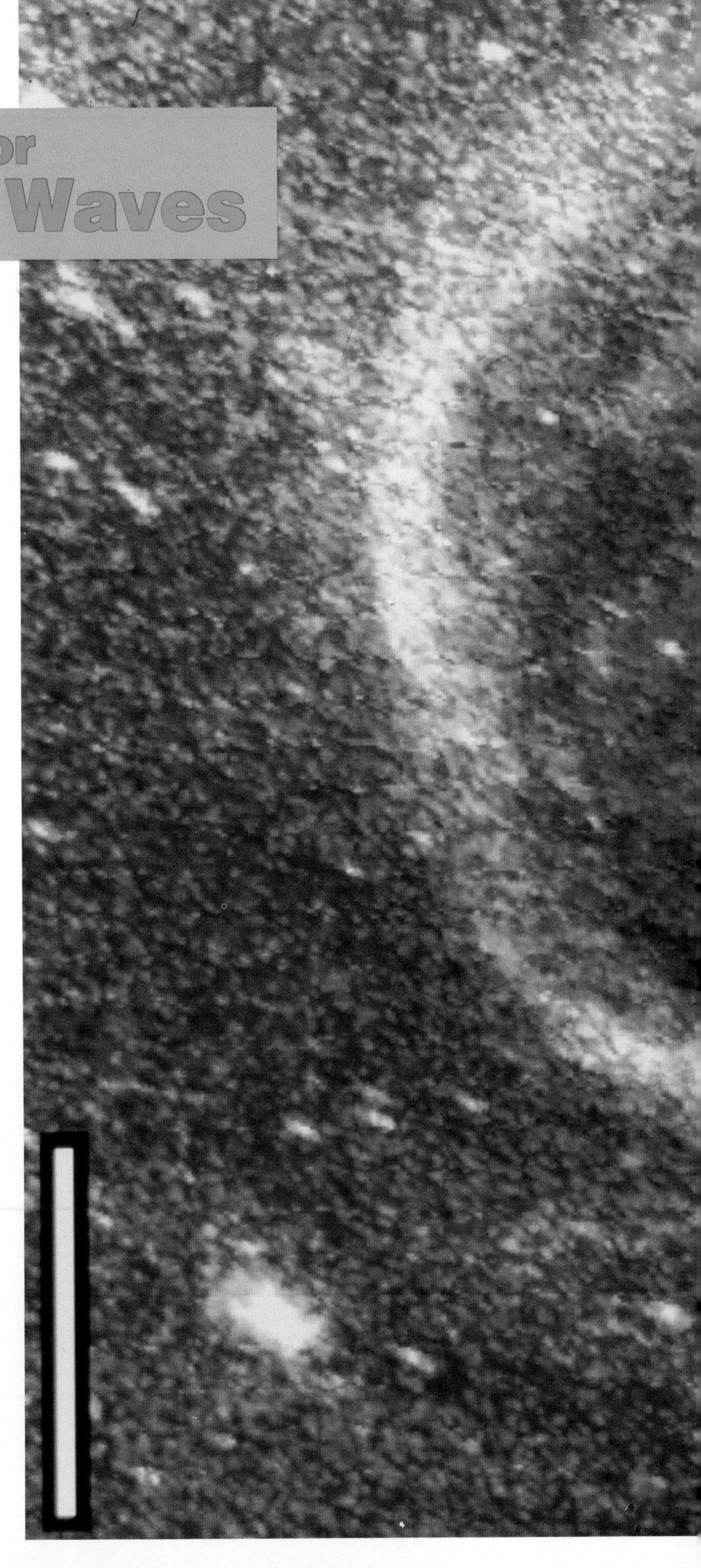

Sophisticated computer image processing has shown that density waves control the evolution of spiral arms in galaxies.
by David J. Eicher

Astronomers using new computer methods and old-fashioned ingenuity have confirmed what they have long suspected: The spiral shapes observed in many galaxies are sustained by density waves.

Astronomer Bruce Elmegreen of IBM's Thomas J. Watson Research Center in Yorktown Heights, New York, announced his findings at the January meeting of the American Astronomical Society in Arlington, Virginia. Elmegreen and his associates used a sophisticated computer program to "stretch" an enhanced photograph of the galaxy M81 in Ursa Major to represent its face-on appearance. Once accomplished, this image allowed the astronomers to see gaps in the galaxy's arms that show evidence of density waves in action.

"If you look at a conventional photograph of a galaxy, your eyes can play tricks on you and connect a spiral that's not really connected," said Elmegreen. "When we looked at the enhanced images we could see right away that these spirals are broken, and the breaks occur at the same place on both arms, preserving the overall symmetry."

Elmegreen reported that his image of M81 and two others of galaxies M51 in Canes Venatici and M100 in Coma Berenices show that density waves prevent a peculiar fate for spiral galaxies. Astronomers have long recognized that something must be sustaining the pinwheel shape of spiral galaxies, which rotate at high speeds around their centers. If the spiral arms simply trailed the rotation of the galaxy's hub, the arms would eventually wind up tightly around the galactic core.

Density waves, moving regions in the galaxy carrying contrasting high and low density pockets of stars, regulate a galaxy's overall shape. When

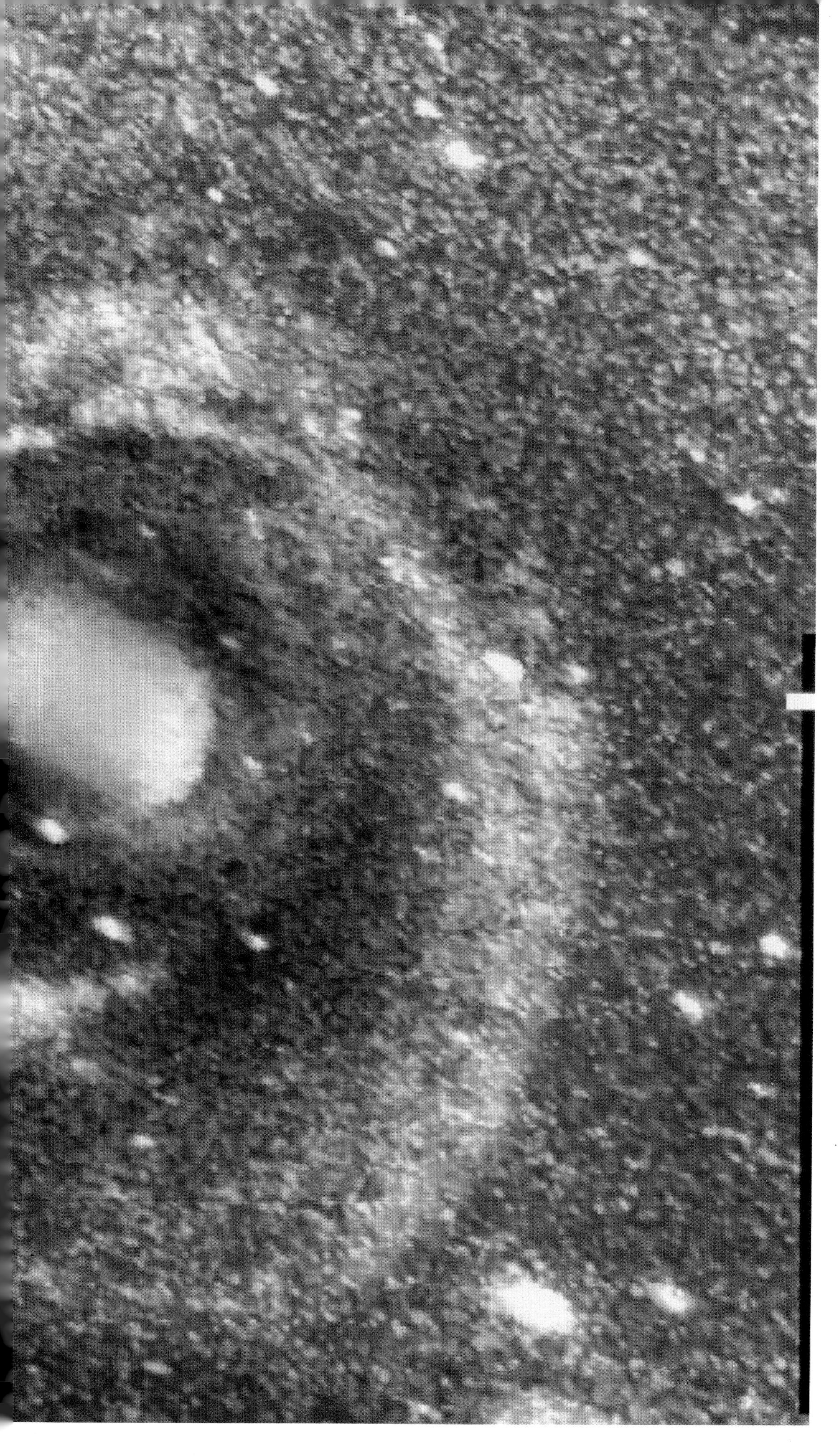

BROKEN SPIRAL ARMS reveal evidence of density waves acting in the galaxy M81 in this computer-processed image. Density waves sustain the spiral patterns in galaxies over long time scales. Photo courtesy Bruce Elmegreen, Thomas J. Watson Research Center.

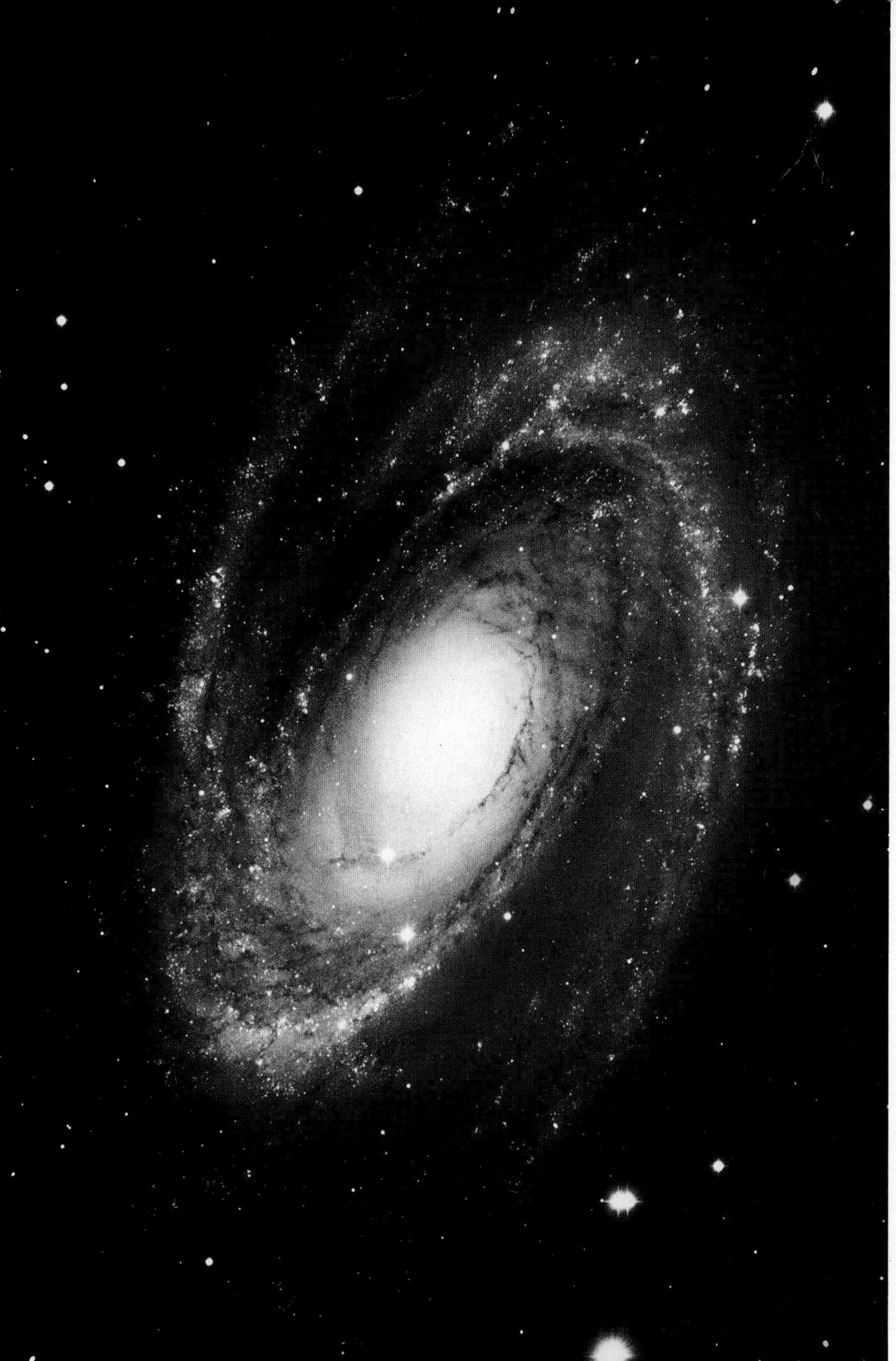

A NORMAL VIEW of M81 fails to show gaps in the galaxy's arms. After computer processing, hundreds or thousands more galaxies may show evidence of density waves. Palomar Mountain Observatory photo.

the dense part of a density wave encounters a star, it causes the star to slow down in its orbit, packing numerous stars together like helpless autos in a traffic jam. The dense pockets of stars reinforce the wave's gravity, thus further bunching additional stars.

The newly observed gaps in the spiral arms indicate that the spiral shapes of galaxies are preserved by two sets of density waves moving in opposite directions. The gaps occur in areas where the crest of one wave meets the valley of another. Stars that pass through these areas are not affected at all; they simply continue along their orbits, leaving gaps in the arms.

The first direct proof of the density wave theory comes twenty-five years after the idea was proposed by astronomers C. C. Lin and Frank H. Shu, based on work by Swedish astrophysicist Bertil Lindblad. The most recent incarnation of the theory appeared in an article in the March 1989 issue of the *Astrophysical Journal,* written by Lin, Giuseppe Bertin, S. A. Lowe, and R. P. Thurstans of the Massachusetts Institute of Technology. The theory predicts that two density waves interfering in this way can result in a spiral pattern that persists for several revolutions of the galaxy, thus avoiding the winding dilemma.

The enhanced images produced by Elmegreen also show small "spurs" of stars emanating from the spiral arms of these galaxies. These spurs allow astronomers to identify the significant regions of resonance where repeated encounters with density waves wildly scatter the orbits of individual stars. By finding these regions the astronomers were also able to determine the speeds at which density waves travel through the galaxies.

Just how density waves get started and what their long-term effect on the organization of stars in spiral galaxies is is not yet known. However, Elmegreen and others are busily working to find the answers to these questions. For now, the astronomers are happy to have confirmed one of the longest standing theories about how spiral galaxies behave. □

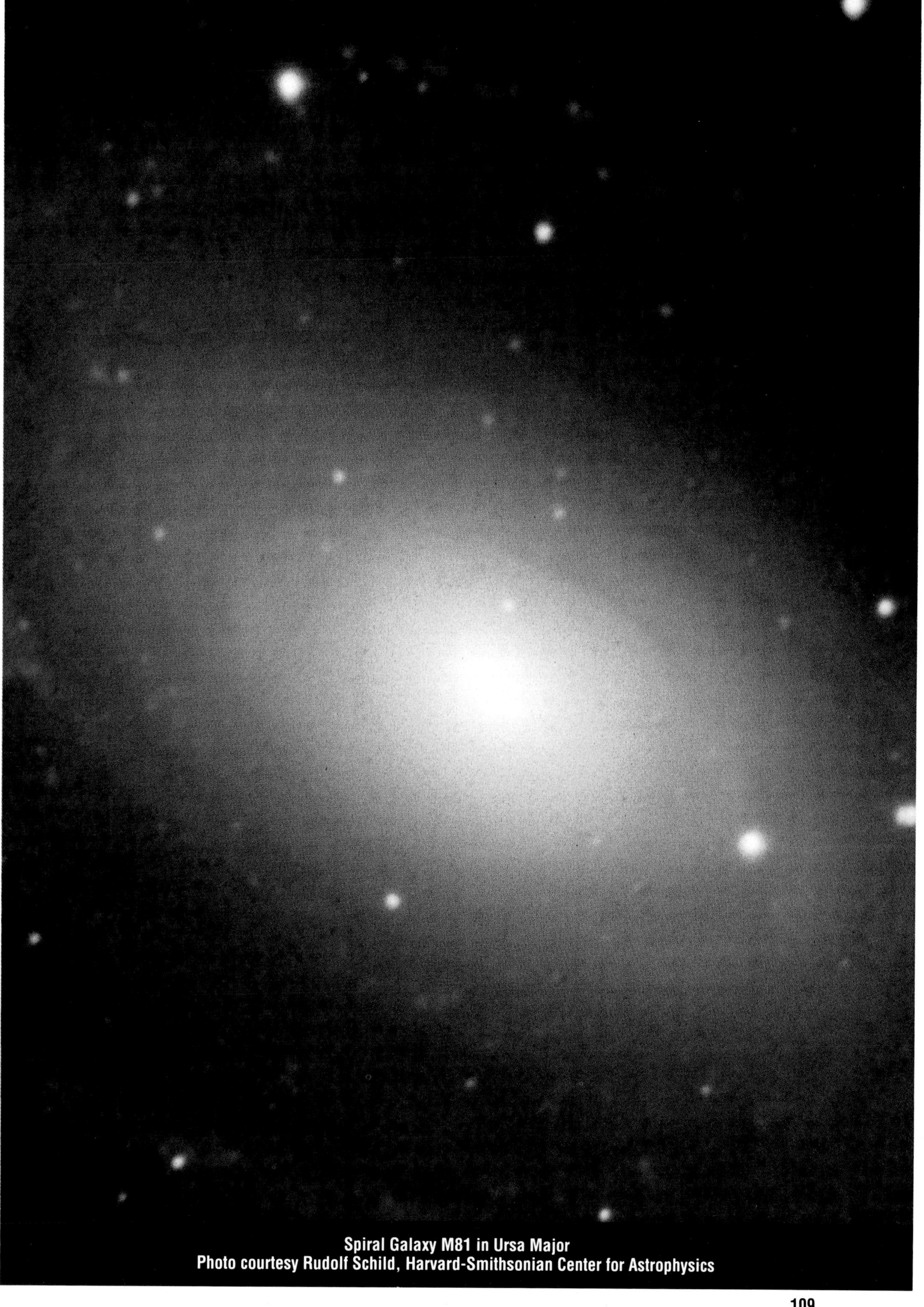

Spiral Galaxy M81 in Ursa Major
Photo courtesy Rudolf Schild, Harvard-Smithsonian Center for Astrophysics

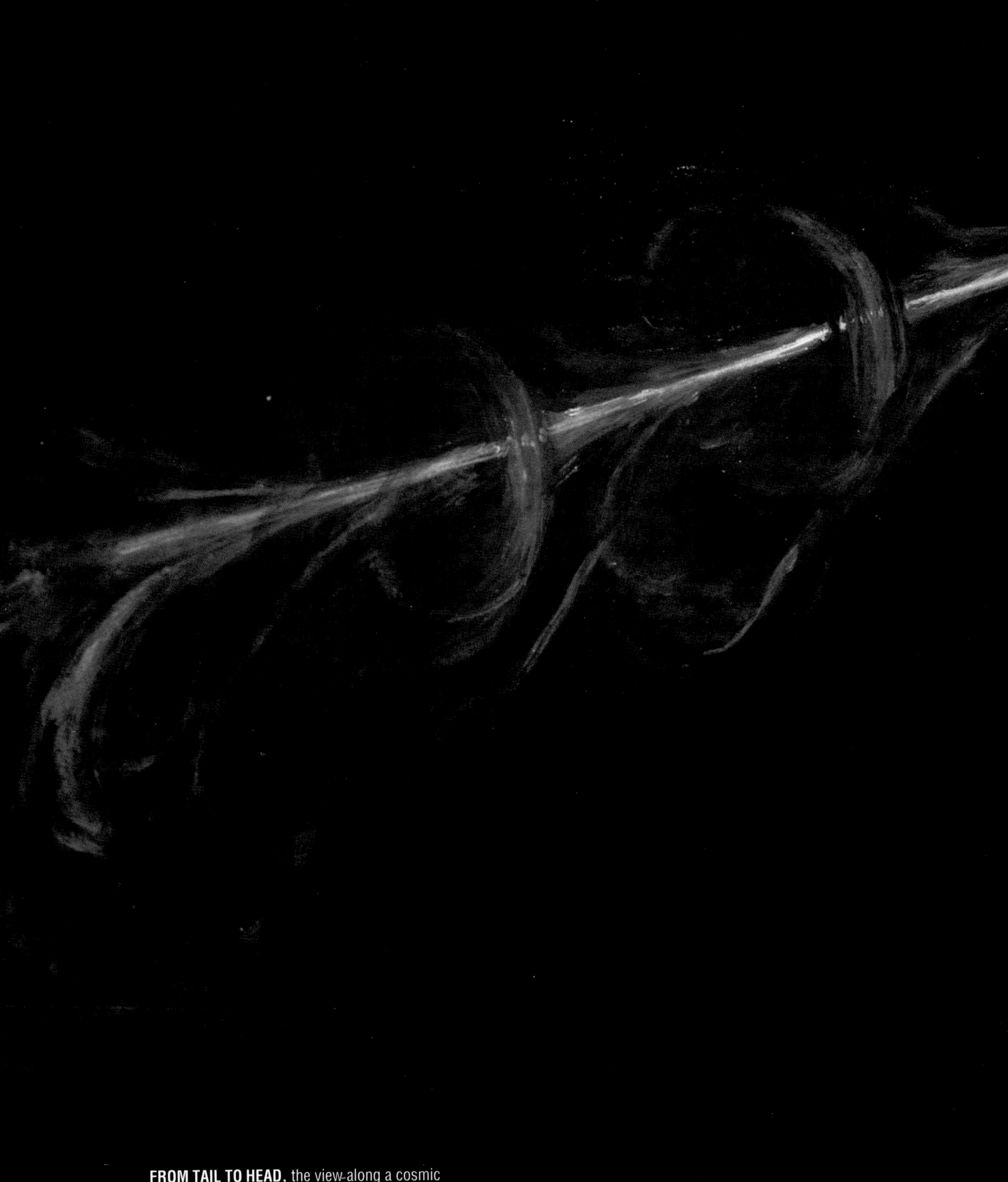

FROM TAIL TO HEAD, the view along a cosmic jet back toward the nucleus of an active galaxy. Painting by Thomas L. Hunt.

Chasing the Monster's Tail: New Views of Cosmic Jets

Radio telescopes and supercomputers are working together to reveal new features in the mysterious jets of gas emanating from the cores of active galaxies and quasars.

by Jack O. Burns

Within the last decade astronomers have been able to view active galaxies and quasars in ever-increasing detail. Whereas only the most obvious features are visible with optical telescopes, data collected by radio telescopes, such as the Very Large Array (VLA) in New Mexico, have allowed astronomers to produce images of these energetic objects at longer, more telling wavelengths.

By far the most intriguing structures seen in some of these images are cosmic jets — streams of high-energy particles that spew out of the nuclei of active galaxies and quasars and stretch for tens of thousands of light-years out to giant diffuse lobes or "tails." Sensitive radio maps of these intriguing features reveal complex knots, filaments, and loops embedded within the jets and lobes. Moreover, they show that cosmic jets come in a variety of different shapes, which some astronomers think may be due, in part, to the interaction of the jet's high-velocity gas with the intergalactic environment.

Just as observing a river from the banks reveals little useful information about the river's source, history, and hydrodynamics, observations of cosmic jets can't tell astronomers what the nature of these structures are or how they form and evolve.

Recent advances in computer software and hardware, however, have allowed astronomers not only to construct very high-contrast images of jets, but to model their behavior as well. This synergistic relationship has opened the door to new and exciting theoretical insights. Using supercomputers, astronomers are now able to numerically simulate the development of radio jets as they propagate from the cores of galaxies into the intergalactic medium. The results of the simulation's calculations are then converted into radio emission images that can be directly compared with VLA maps of the real thing. For the first time, astronomers can test theoretical models directly against real radio data with comparable resolution and contrast.

Feeding the Monster

Only a few percent of all galaxies and quasars are sources of extended radio emission. Yet these objects intrigue astronomers because of their structural beauty, their power, and the exotic mechanism that fuels the radio sources in their cores. These relatively rare beasts spawn giant lobes of radio emission with enough energy to equal or exceed that from the hundreds of billions of stars within the galaxy or quasar. The engines that drive these radio sources are probably giant black holes with masses ranging from 10 million to a billion suns.

A black hole is the favorite candidate for the "monster" at the core of radio galaxies and quasars because it is highly efficient in converting matter into energy. The strong gravitational field of a black hole accelerates and heats the gas that falls onto a flattened accretion disk swirling around the rotating hole. The energy and high-speed electrons, in the form of jets, are then redirected back out along the magnetic field lines of the black hole's north and south magnetic poles, a direction that is perpendicular to the accretion disk. The jets are simply blown out of the galaxy/quasar core along the direction of least resistance by the heat and strong electromagnetic dynamo action of the accretion disk.

Astronomers still can't explain why so few galaxies and quasars are radio sources. Recent optical observations, however, indicate that a fairly high percentage of radio galaxies and radio quasars are "disturbed" in appearance — the optical images show filaments, plumes, or tidal tails. These features are indicative of recent gravitational interactions with nearby companion galaxies. Such tidal encounters could seriously stir up the inner cores of galaxies, forcing fresh gas or disrupted stars onto the accretion disks of black holes. This mechanism may provide the fuel needed to power the radio source.

Classes of Radio Sources

In the mid-1970s, B. L. Fanaroff and J. M. Riley from Cambridge University recognized that extragalactic radio sources could be divided into two classes according to their power, or energy output, and structure.

Class I radio sources are weak in terms of total radio luminosity, but their jets have complex forms and structures, something that extragalactic astronomers refer to as morphology. These sources are almost always associated with galaxies rather than quasars. Generally, Class I sources are elongated and their radio emission diminishes in intensity with distance from the radio galaxy. They often have two jets that emerge in opposite directions from the galaxy core.

For some Class I radio galaxies that are located in clusters, the jets are bent into spectacular U-shaped structures that eventually flare into diffuse "tails." These galaxies are referred to as head-tail sources — the head being the galaxy, the tails being the radio-emitting lobes fed by jets. The bend in the jet is produced by the radio galaxy's rapid motion (sometimes thousands of kilometers per second) through a hot, dense intracluster medium, whose presence is known via its x-ray emission. The tails are swept back behind the radio galaxy by the dynamic pressure imposed by this medium, much like the pressure of the wind against cigarette smoke blown from a moving automobile.

Class II sources are the most powerful. They generally are referred to as "classical doubles," because a compact core, coincident with the nucleus of the galaxy or quasar, lies midway between two lobes of strong radio emission. The radio surface brightness peaks at "hot spots" located at the extremities of the lobes. A single jet is most often found emanating from the core of Class II sources "feeding" into what is generally the brightest and most compact hot spot in one of the lobes.

The Structure of Classical Doubles

For nearly a decade, astronomers have been puzzled by the fact that classical doubles exhibit only one-sided jets. One model that explains this jet asymmetry assumes that each of the lobes is powered by a jet, but we can only see the jet that is moving in our direction. The gas or plasma in the jet is presumed to be moving very close to the speed of light. The special theory of relativity predicts that the jet pointing toward us will be boosted in intensity and the counterjet pointing away will be diminished by the same factor. Thus, relativistic effects cause one jet to appear much brighter than the other. In principle, if astronomers could produce even more sensitive maps, we would begin to see the dimmer counterjet. Recent high-contrast VLA maps are revealing possible pieces but no complete counterjets yet.

In 1974, Roger Blandford and Martin Rees from Cambridge University proposed a "twin-exhaust" jet

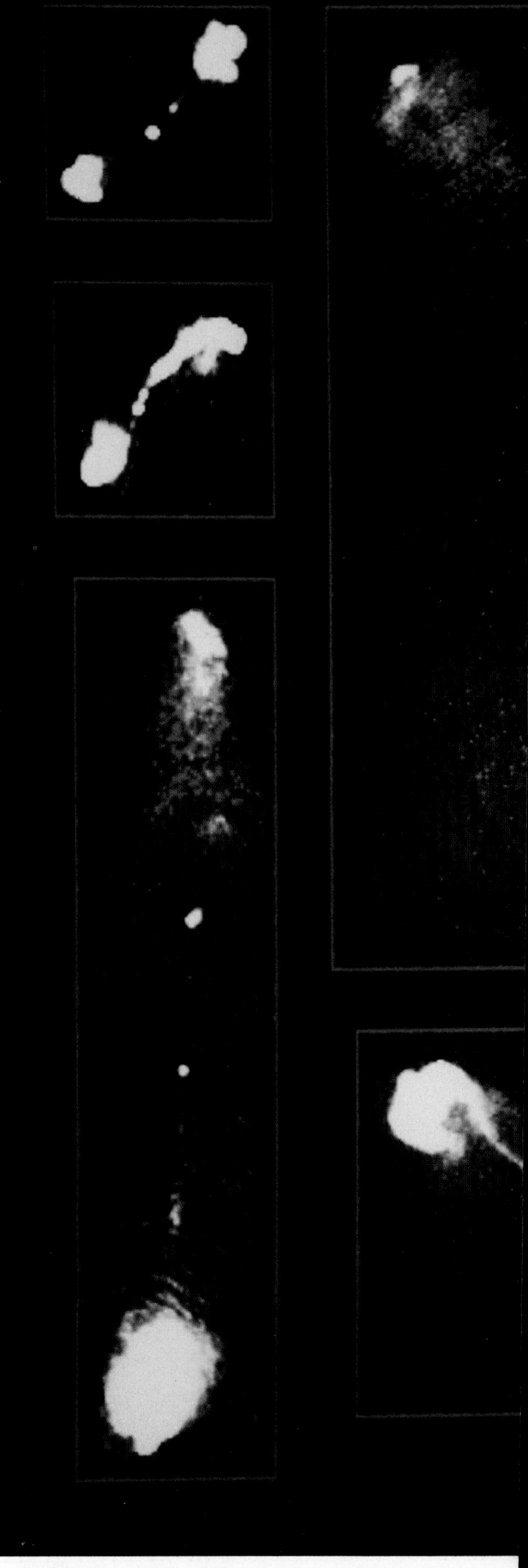

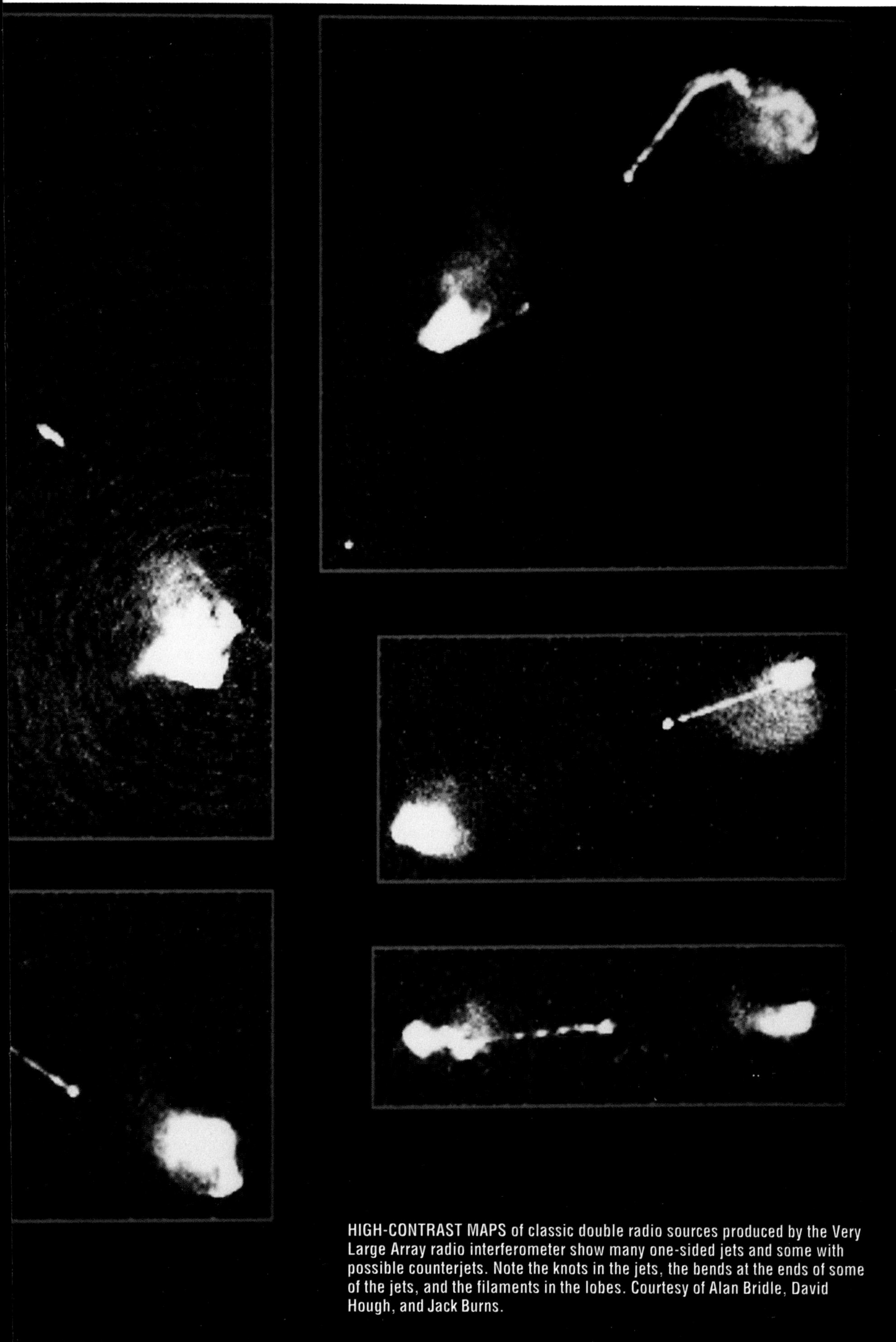

HIGH-CONTRAST MAPS of classic double radio sources produced by the Very Large Array radio interferometer show many one-sided jets and some with possible counterjets. Note the knots in the jets, the bends at the ends of some of the jets, and the filaments in the lobes. Courtesy of Alan Bridle, David Hough, and Jack Burns.

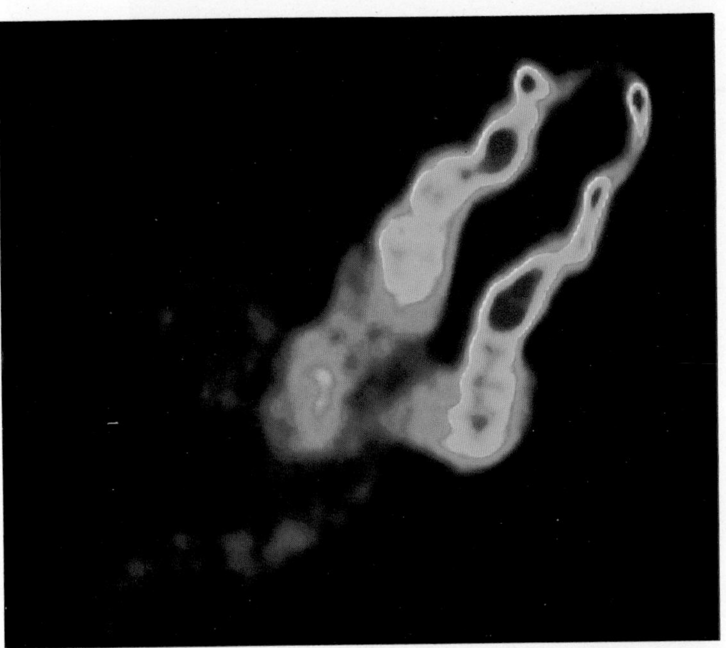

HEAD-TAIL RADIO SOURCE 3C 129 exhibits a distinctive U-shaped structure. The galaxy ("head") is located near the lower left at the base of the jets. Two jets emerge from the galaxy core and are blown back into diffuse "tails" by the rapid motion of the galaxy through the intergalactic medium. Courtesy of Larry Rudnick and Jack Burns.

model to explain the basic morphology of classical double sources. (This model is particularly noteworthy since it predicted the existence of jets before they were discovered with radio interferometers.) Blandford and Rees speculated that thin, narrowly-focused, fluid-like beams or jets born near the central galaxy/quasar engine are driven outward by the kinetic energy of the plasma beam — much like the exhaust from a supersonic aircraft. At the outer, leading end of the jet, called the working surface, a strong shock forms as the jet's plasma pushes into the surrounding medium, causing the jet material to rebound toward the galaxy by "ram" pressure. Blandford and Rees interpreted the hot spots in classical doubles as the shocks at the working surface of the jets, and the lobes as the diffuse jet material blown back by the intergalactic wind.

By the early 1980s, supercomputer simulations by astronomers Michael Norman, Karl-Heinz Winkler, and Larry Smarr, using a Cray computer at the Max Planck Institute in West Germany, verified and significantly expanded upon the Blandford-Rees model. These and subsequent simulations used the basic equations of fluid dynamics to follow the evolution of a cylindrical jet out of a galaxy core and into the intergalactic medium. From these studies, astronomers have concluded that two key parameters appear to affect the structure of the classical doubles: the velocity of the jet in terms of its Mach number (a measure of how many times the jet velocity exceeds the speed of sound in the jet gas) and the ratio of the jet's gas density to that of the intergalactic medium.

It appears that the jets in classical doubles are highly supersonic, with Mach numbers of 10 or greater, and that the jet plasma is light (that is, less dense than the surrounding medium). In the weaker but more complicated Class I sources, the radio plasma also appears to be less dense than the surrounding gas, but the jets are only mildly supersonic (Mach numbers of 2 to 5) and the tails are subsonic. Supercomputer simulations of such jets reproduce both their overall shape and many of the detailed structures seen in recent VLA maps.

Numerical Observations

Over the past five years, the creation of several national supercomputing centers by the National Science Foundation has allowed these types of simulations to be run in the United States. Michael Norman, now at the National Center for Supercomputing Applications at the University of Illinois, and I, along with students and postdoctoral fellows, have formed a unique collaboration to study both the observational and theoretical aspects of radio jets using Cray computers.

But a powerful supercomputer alone is not enough to model these extended radio sources. The computer must be programmed with the basic laws of physics, and sophisticated schemes must be developed that allow jets to evolve in time within the confines of these laws. The computer program that we run at the University of Illinois — called ZEUS — was written over a period of more than a decade. The basic design of the code began with the supercomputer pioneer James Wilson at the Lawrence Livermore Laboratory. For his Ph.D. dissertation Norman adapted the program to investigate cloud collapse and star formation. Later at the Max Planck Institute in West Germany, Norman, Winkler, and Smarr expanded the program to model the extragalactic jets that had recently been observed in great detail with the VLA. Students and postdocs have continued to rewrite and advance ZEUS, which now has more than 25,000 lines of computer code.

A hydrodynamics program such as ZEUS begins with the equations for a compressible gas or fluid, which were originally derived by the brilliant Swiss mathematician Leonhard Euler in 1755. It took the invention of sophisticated, large-memory computers, however, before the full potential of these equations could be exploited. ZEUS calculates the time-dependent flow of a cylindrical jet on a numerical grid. At each time step, ZEUS updates the hydrodynamic variables of density, pressure, and entropy (a measure of disorder in the system) according to Euler's equations. By solving Euler's equations in this fashion, astronomers and physicists can reproduce the complex and chaotic structures that are viewed in nature — such as the turbulence seen behind a motorboat and the instabilities in a wiggling firehose. The complex, nonlinear process inherent in Euler's equations can only be extracted and tracked using numerical techniques and a large computer.

For his Ph.D. dissertation, David Clarke significantly extended ZEUS by including magnetic fields in the calculations. This allowed us, for the first time, to produce maps of radio emission similar to those obtained by the VLA. (The radio emission observed from jet sources in galaxies and quasars is generated by fast electrons spiraling in magnetic fields, a process termed synchrotron radiation.) These "numerical observations" have placed supercomputer models onto the same footing as VLA observations where we can directly compare emissions, and thereby probe the underlying physics in unprecedented detail.

One new result from our numerical observations is that jets with very strong magnetic fields tend to produce lobeless sources. Magnetic forces accelerate jet plasma beyond the head of a hydrodynamic jet into a thin "nose cone" structure. Such sources do not have flow-back material or cocoons. There are a few quasars, such as 3C 273, that appear to have nose cones without cocoons, but such structures are rare. My colleagues and I have concluded that magnetic fields do not play a dominant role in forming the structure of classical double sources — a finding that was only possible using the supercomputer.

Born-Again Radio Jets

Astronomers have known for more than thirty years that the luminosities of compact cores of radio sources vary in time. More recently, Very Long Baseline Interferometry (VLBI) has demonstrated that new jet knots often appear after a flare in the intensity of an active galaxy or quasar nucleus has been observed. Apparently, the plasma output of the engines in such sources is not constant in time. There is further evidence of variable outputs from galaxies in sources such as 3C 219 where partial jets are observed.

Nearly all previous models of classical doubles assumed that jets were continuously turned on. The variability of the galaxy or quasar core may have important consequences for jet and lobe structures, however. As another example of the insight that can be gained from numerical simulations, David Clarke and I recently ran a series of supercomputer models in which we "turn on" a jet, turn it off for an equal period of time, then turn it on again. The results show that a "reborn" jet propagates much more quickly (down the evacuated channel left by the previous jet) than does the original jet. Such reborn jets catch up quickly with the previous working surface, and this could explain why measured velocities of some newly-seen knots in VLBI images of jets are moving at different, sometimes greater, velocities. We also found that when the simulated jet is turned off, the lobe and hot spot expand and cool. This agrees with observations in which an active jet in a classical double is seen on the side with the more compact lobe hot spot.

Most importantly, the time-variability problem illustrates the control and flexibility that astronomers now have in studying jets. Because of the long lifetime of radio sources (hundreds of millions of years), we see only a brief moment in the total life of a jet with our radio telescopes. It is difficult, then, to understand the long evolutionary history of such sources. With a supercomputer, however, we can collapse this timeline, watch how sources evolve under different circumstances, and attempt to understand the history behind the variety of sources that we observe with the VLA. Supercomputers allow us to control both time and space.

Disruption of Radio Jets

Now that some progress has been made toward understanding the jets in classical doubles, we have begun to turn our attention toward understanding the more complex jet structures associated with Class I sources. There is a subset of radio galaxies, termed wide-angle tailed sources, that lie at the break between Class I and Class II. Aileen O'Donoghue of St. Lawrence University has recently produced some exquisite VLA maps of these sources identified with optically bright galaxies at

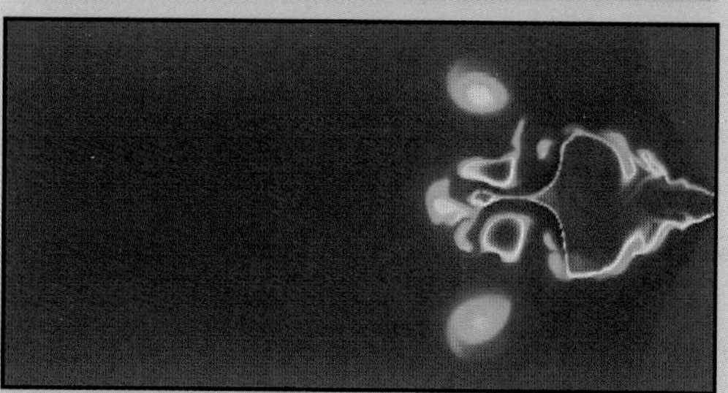

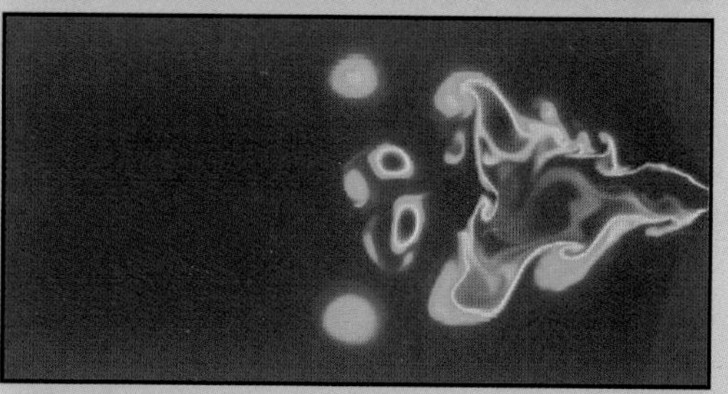

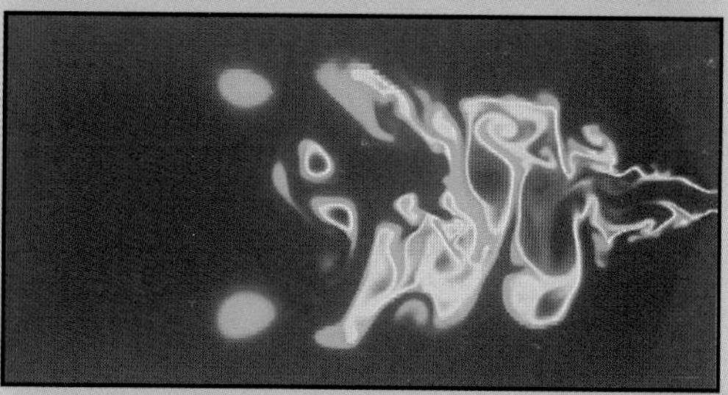

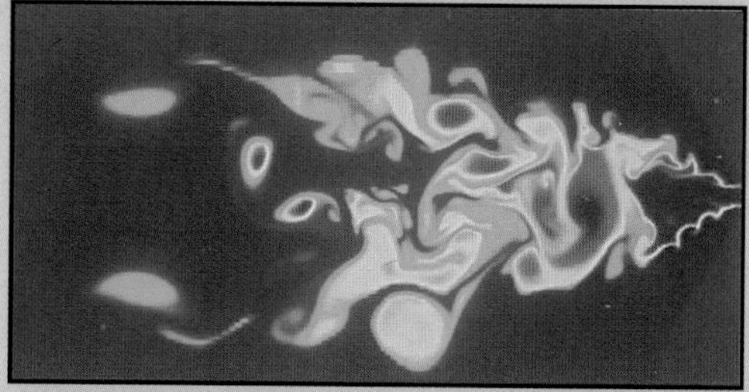

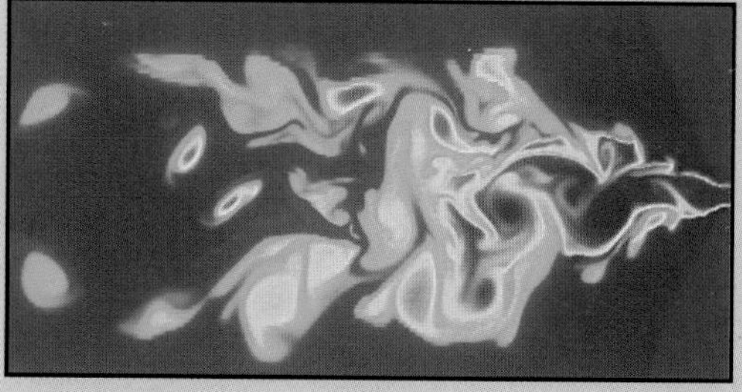

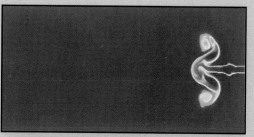

A JET DISRUPTS when it encounters the dense intergalactic medium, according to this computer simulation. The jet inlet is on the right and propagates to the left. Areas of greatest density are colored blue; least dense areas are red. Courtesy of Jack Burns.

THE RADIO LOBES of Centaurus A stretch over 20,000 light-years from the nucleus of the galaxy (superimposed). Centaurus A is the closest example of an active galaxy and a Class I radio source. Radio image courtesy of David Clarke and Jack Burns. Photo of NGC 5128 from the Anglo-Australian Telescope Board.

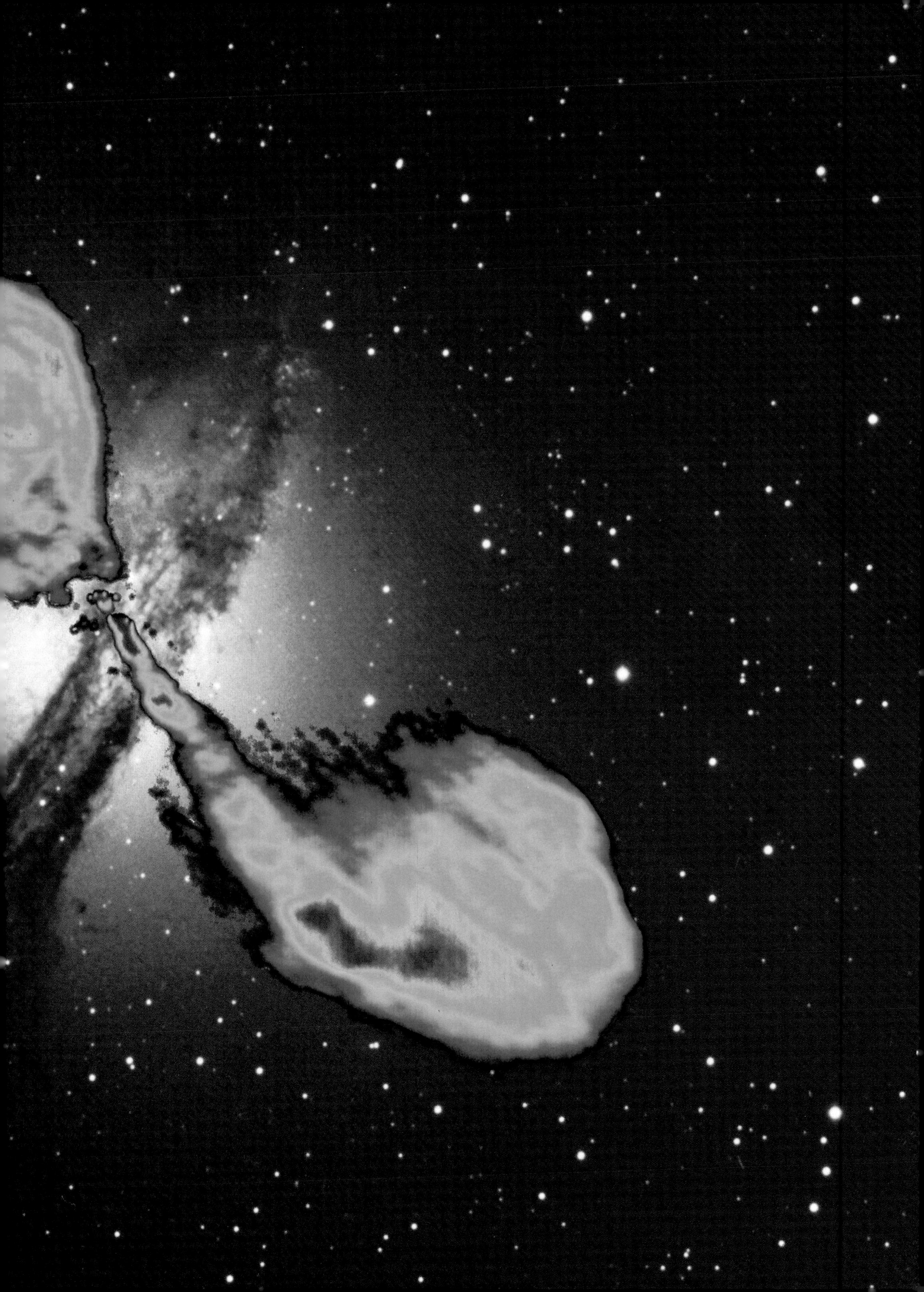

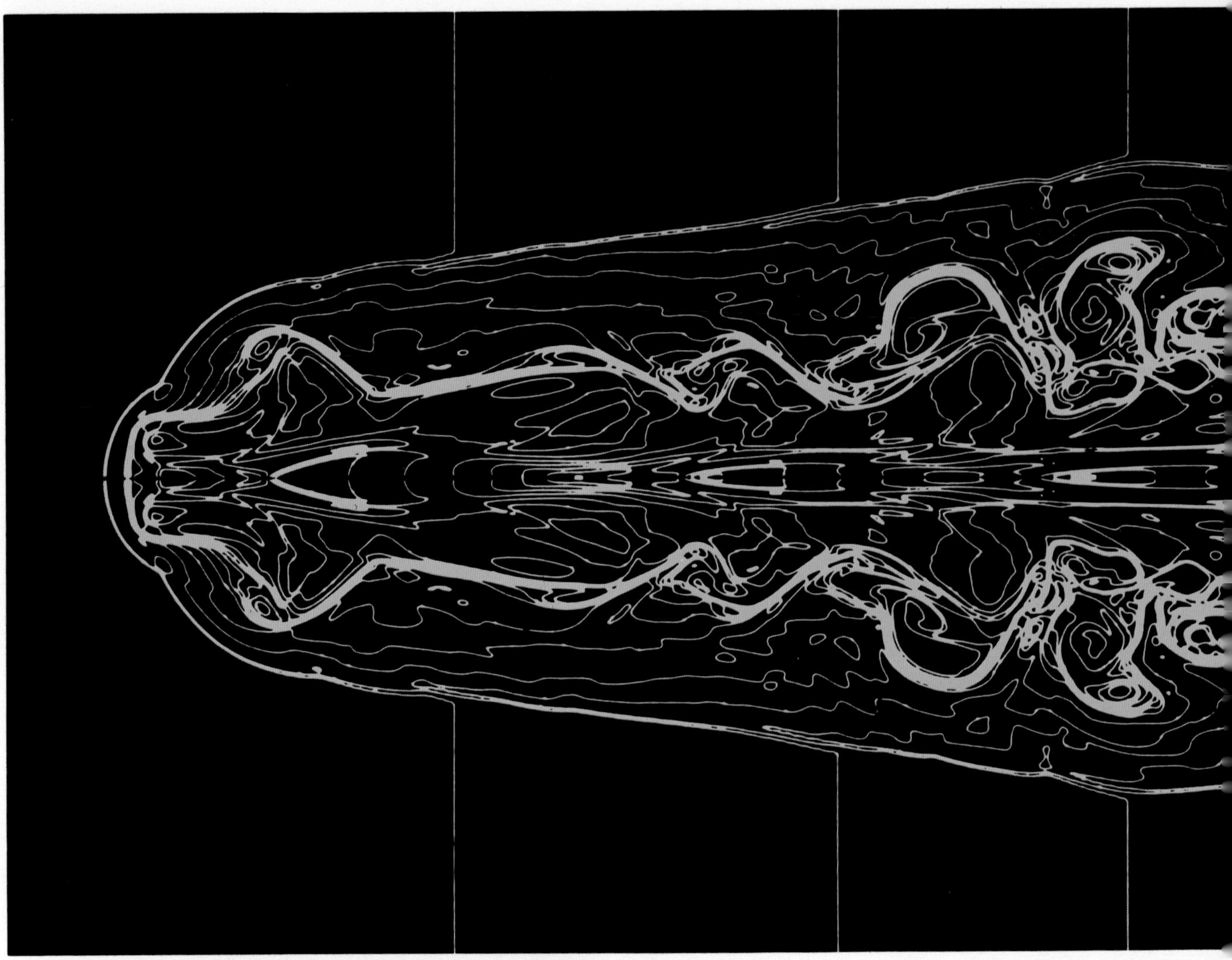

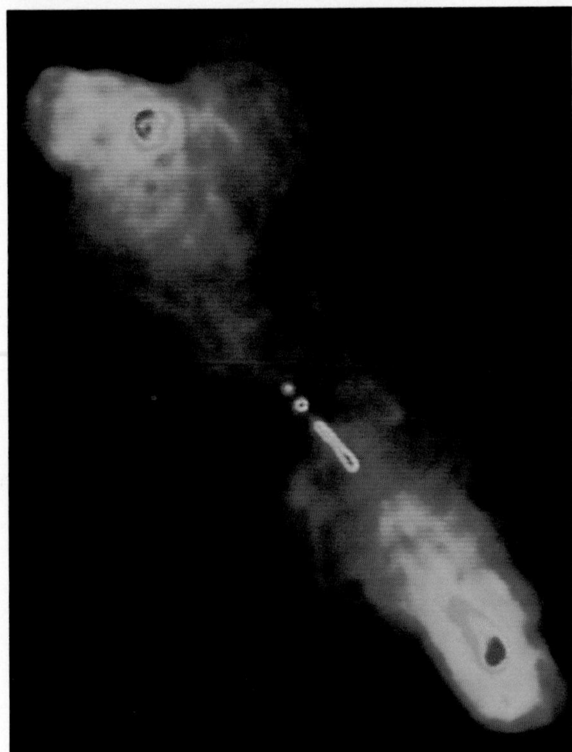

SUPERSONIC JETS, like those emerging from 3C 219 (left) and M87 (above right), consist of complex structures. Pushed out ahead of the jet's working surface is the bow shock. Behind the bow shock is the "contact discontinuity," a boundary that separates the shocked intergalactic material from a cocoon of plasma, which is blown back down the outer edge of the jet by the intergalactic wind. Material in the cocoon is subsonic and highly turbulent. The eddies exert a force back upon the jet that produces crisscross shocks, which may in turn produce bright, compact knots, like those in M87. Schematic courtesy of Jack Burns. Photo of 3C 219 courtesy of David Clarke. Photo of M87 courtesy of NRAO/AUI.

the centers of clusters. Wide-angle tailed sources have twin jets that are very straight. They travel outward some 30,000 to 100,000 light-years from the galaxy's nucleus, which is still within the galactic halo, where they suddenly disrupt, flare, and bend. The origin of this sudden disruption is a mystery.

Michael Norman, postdoctoral fellow Martin Sulkanen, and I think that a "pressure wall" may exist at the interface between the interstellar medium of the galaxy and the intergalactic medium. Such a pressure wall may be produced by a supersonic "wind" that blows radially outward from the core of the active galaxy. When the wind strikes the stagnant intergalactic gas, a shock is produced. We theorize that if a mildly supersonic jet

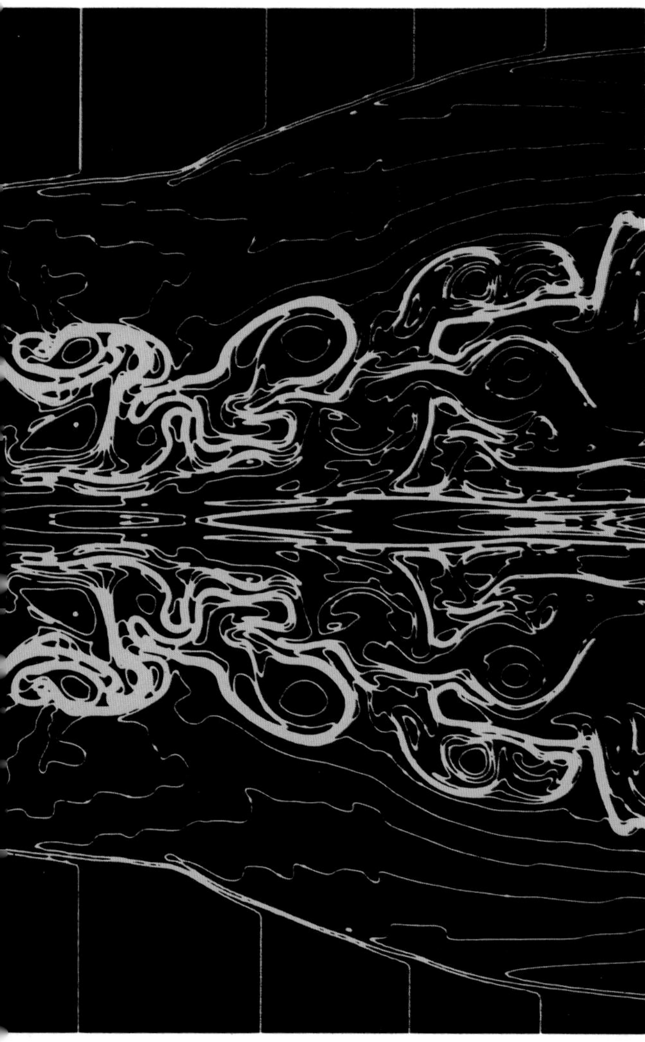

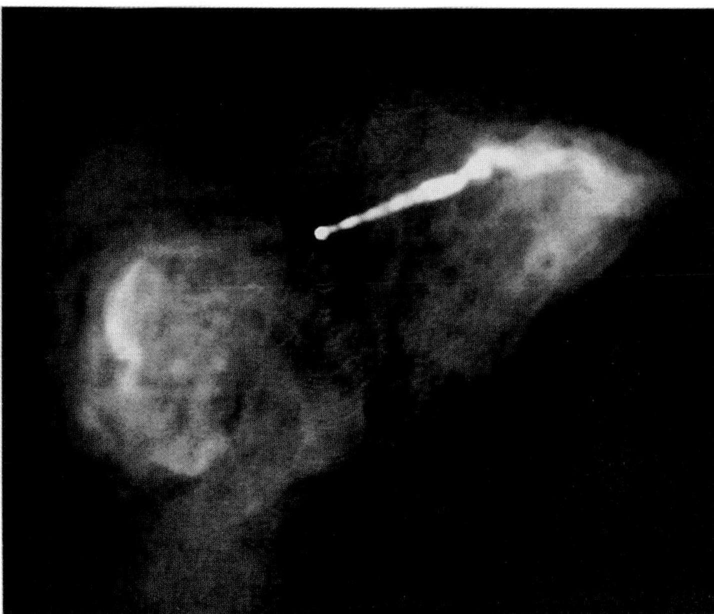

passes through such a shock, the jet won't be able to readjust to the sudden change in external pressure and will disrupt.

Our numerical simulations have verified this theory. When the jet strikes the external shock, an internal Mach disk, or a planar shock, is formed in the jet. This causes the jet material to decelerate and become subsonic. The laws of fluid dynamics tell us that the subsonic jet must expand if its pressure is to balance with the pressure of the surrounding medium. Thus, the jet is seen to flare abruptly as in the VLA observations. Furthermore, because of its lower velocity, a subsonic plasma is highly susceptible to turbulence. Thus, the expanded jet or tail is very chaotic, much like smoke rising from a chimney.

Since our simulations reproduce the basic observations so well, we think there may be a previously unrecognized gaseous component in elliptical galaxies. These galaxy pressure walls are recognized as a result of new VLA radio observations coupled with supercomputer simulations.

The research doesn't stop there, however. Jun-Hui Zhao, a graduate student at the University of New Mexico, has recently demonstrated that an external shock is not the only mechanism that will disrupt a jet. X-ray observations of galaxy clusters with dominant central galaxies often show evidence that gas in the cluster is cooling and falling in toward the center of the cluster. The gas cools by radiating x-rays and is driven toward the cluster's center by pressure from the hotter gas in the outskirts of the cluster. The particular profile of intergalactic gas pressure in these "transonic" cooling flows — the pressure drops toward the center of the cluster as the flow becomes supersonic — also appears to disrupt jets according to our numerical simulations. Again, this agrees with observations in which clusters with cooling flows have jets that are small in length with apparent disruptions near the cluster center.

The work to date has shown that numerical simulations can be a powerful tool for understanding extragalactic radio sources. However, these previous numerical simulations have been constrained to two dimensions due to limited memory in the supercomputers. With the arrival at the University of Illinois of the Cray 2 computer, which has more memory capacity than previous supercomputers, we are now expanding our simulations into three dimensions. This will allow us to tackle a broader range of problems, such as the bending of radio jets in head-tail sources.

We have entered a new era in astronomy in which computers have given us the power needed to image radio emissions from galaxies and quasars with unprecedented clarity. At the same time, supercomputers have permitted us to study the evolutionary history and underlying physics of extragalactic radio sources with numerical simulations. In the exciting years to come, powerful groundbased radio telescopes, the *Hubble Space Telescope*, and computers will allow astronomers to view and model the cosmos with a bold new perspective, yielding fresh insights into the structure and origin of cosmic jets. □

Jack O. Burns is head of the astronomy department and professor of astronomy at New Mexico State University. He is also an adjunct researcher at the National Radio Astronomy Observatory and has been a long-term visiting scientist at the University of Illinois' National Center for Supercomputing Applications.

©M.Carroll
91

The Emerging Picture of Quasars

Radio, infrared, ultraviolet, and x-ray radiation from quasars shows that they are supermassive black holes surrounded by a thick disk of dust and gas.

by Belinda J. Wilkes

Quasars are the brightest objects in the universe. From a volume perhaps no larger than our solar system, as much light as a trillion Suns pours forth. These enigmatic objects have continually fascinated, puzzled, and surprised astronomers during the 28 years since they were discovered. Recently, observations at wavelengths across the electromagnetic spectrum and advances in instrumentation have given us a more complete view of quasars and a much better understanding of what they are and how they work.

A Puzzling Discovery

As is often the case in astronomy, our first view of quasars was misleading. In 1963 astronomers discovered two strange radio sources — unusual in that they were strong, small, and varied rapidly — during the course of the 3rd Cambridge (3C) radio survey. Optical photographs of the regions surrounding both 3C 48 and 3C 273, as they were called, showed only an innocent-looking blue star. Because no stars known at that time could emit so much radio radiation, astronomers seemed to have a new and very unusual kind of star on their hands.

Optical spectra of the objects, however, taken to find out what kind of star they were, looked nothing like that of a star. Strong, broad emission lines, unlike any seen previously, dominated the spectra, instead of the usual narrow absorption lines. After puzzling over this spectrum for a while, Maarten Schmidt of Caltech realized that the emission lines were the familiar Balmer series of hydrogen lines shifted to longer (redder) wavelengths. The amount of this redshift implied that the stars were moving away from us at high velocities (for 3C 273, 16 percent that of light). This velocity was much faster than that of any galaxy known at that time and seemed unlikely for a star!

Astronomers developed two main theories to explain the high velocities of quasars. In the ejecta model, strong explosions eject quasars from nearby stars or galaxies. In the cosmological model, quasars are simply very distant objects taking part in the general expansion of the universe. (This is called the cosmological model since cosmology is the study of the universe itself.) Astronomers already knew from galaxy observations that objects at greater distances from Earth appear to be moving away at greater speeds. In the cosmological model, the quasars simply extend this relation to greater distances (about 3 billion light-years for 3C 273).

The large distances implied by the second alternative means that quasars must be extraordinarily bright to be seen from Earth. Not only this, but the short timescale of the radio variability implies that some quasars are only about 1 light-day in diameter, or roughly the size of our solar system. In 1963 no mechanism was known that could generate so much energy in such a small volume, and as a result astronomers were generally reluctant to accept the cosmological picture of quasars. Regardless of the correct model, the name "quasi-stellar radio source," since they look like a star but are in fact very different, was born and the variants "quasar" and "QSO" are the names generally used today.

The search was on! As astronomers looked for objects with matching radio properties, they found more and more quasars. However, they also found many similar objects that emitted little or no radio radiation. In fact radio emission turned out to be the exception rather than the rule with quasars — only about 10 percent of all quasars are strong radio emitters, roughly the same proportion as in galaxies.

The next advance came when astronomers examining detailed photographs found that many quasars had a fuzzy appearance around the bright central object. The spectrum of the fuzz appeared similar to that of a galaxy at the same redshift as the quasar. Thus quasars seemed to lie in the central regions of galaxies but, because they were so much brighter than their host galaxies, fuzz could only be seen in those closest to Earth. They are also frequently situated among groups of galaxies at similar redshifts.

The evidence linking quasars to galaxies supported the cosmological model for quasars. At the same time, no quasars had been found with blueshifted emission lines indicating that they were moving toward us rather than away. Because the

BRIGHTER THAN A TRILLION SUNS, a quasar's central engine dominates the view from the outskirts of its host galaxy. Illustration by Michael Carroll.

SPECTRAL OBSERVATIONS

X-ray

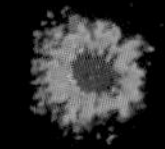

Ultraviolet

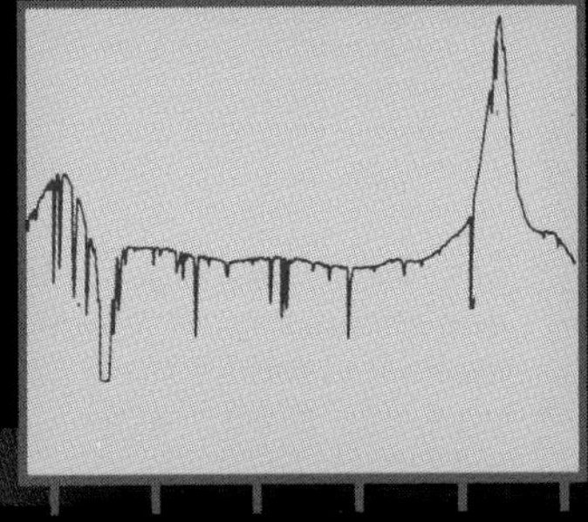

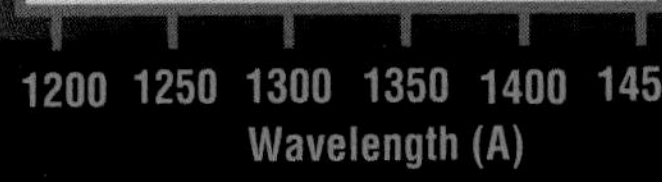

Detailed ultraviolet spectra of quasars are now possible with the Hubble Space Telescope.

Optical

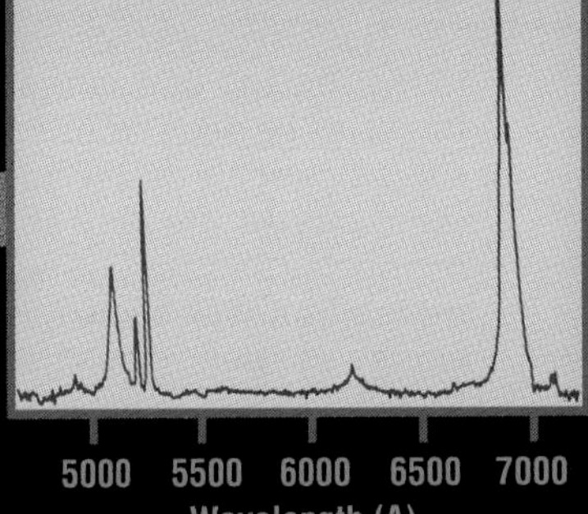

A typical quasar's optical spectrum shows the strong, broad emission lines that characterize quasars.

Infrared

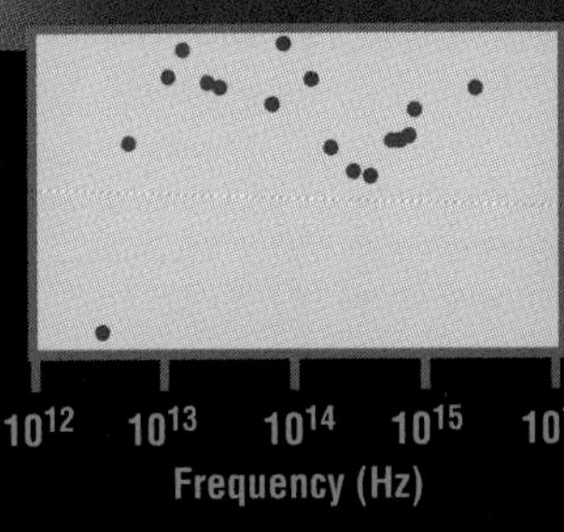

Quasars often show irregular, bumpy shapes in the infrared region of the spectrum.

Radio

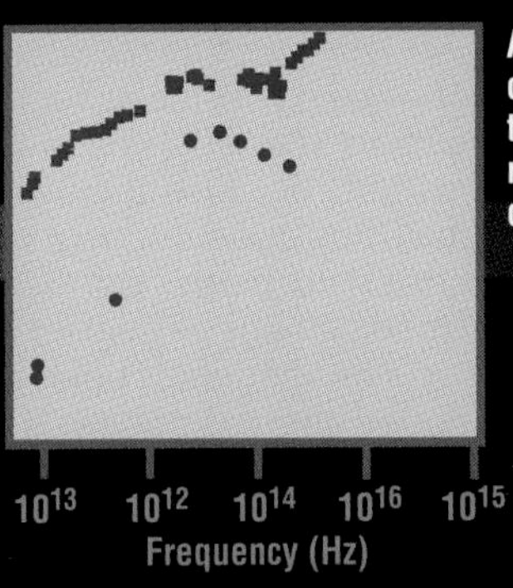

A radio-loud quasar (upper curve) is often several thousand times brighter in radio waves than a radio-quiet quasar (lower curve).

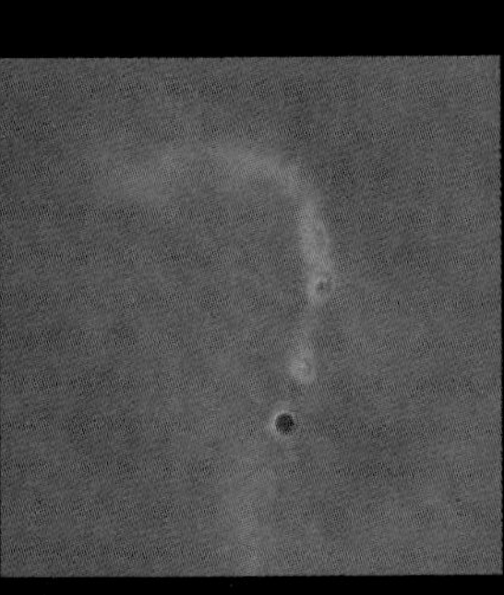

Astronomers study all the radiation emitted by quasars — from radio waves to x-rays — to learn what powers these exotic objects.

ejecta model would have about half of the quasars ejected toward us and half away from us, the balance swung in favor of the cosmological model: A quasar's redshift is due to the expansion of the universe and its recessional velocity is proportional to its distance from Earth. Some 4,000 quasars are now known and more are being discovered each day using a variety of techniques.

The Energy Source

The central problem posed by the cosmological model is how quasars generate enough energy to be seen from Earth. The most efficient way to generate energy is through the gravitational pull of a physically small but very massive and dense object, such as a neutron star or black hole. Matter caught in such a gravitational field spirals in toward the neutron star or black hole, forming a rapidly spinning disk, or accretion disk, around the object. As the matter falls inward, about 10 percent of its mass is converted into energy and radiated as light. This type of infall and energy release is known to occur in binary star systems, where the dense companion object has a mass of about 1 to 6 times that of the Sun. However, the typical amount of light generated by a quasar — some 100 billion to 10 trillion times that of the Sun — requires a central black hole of 100 million times the mass of the Sun. Alternatively, some 100 million neutron stars could be the source, but this seems much less probable.

Knowing a way to generate this much energy solves only part of the problem of understanding quasars. We must also show that our model can produce the kind of energy we observe. A black hole and surrounding accretion disk are certainly small and powerful enough to explain quasars' observed sizes and brightnesses, but what else do we know about them?

Quasars emit roughly equal amounts of energy across the electromagnetic spectrum, from infrared to x-ray and possibly gamma-ray frequencies, and roughly 10 percent also emit strong radio radiation. The multi-frequency nature of this emission has long hampered quasar studies. Stars, which emit most of their energy in optical light, have been fully studied and characterized using optical observations alone. In contrast, quasars emit a wide range of radiation, and we need to observe at all frequencies to completely define their energy distributions — the way their energy is distributed across the spectrum. This requires the use of several different telescopes (infrared, optical, ultraviolet, and x-ray) and observing techniques. Because quasars vary significantly on timescales of a few years, all these data must be collected within about a month to ensure that variability does not affect the results.

Although different quasars exhibit different energy distributions, there appears to be an underlying simplicity to the apparent phenomena. Over the past five years or so, we've learned from multi-frequency observations that the energy distributions can be characterized in terms of a number of different components that dominate the emission in different frequency bands and whose relative strengths change from quasar to quasar. By studying the range of shapes and strengths of each component in comparison with what we expect, we can test our models of how the energy is generated.

An obvious place to start is with radio emission because we already know that there are two classes of quasar: radio-loud and radio-quiet. Surprisingly, the energy distributions of the two classes are very similar in the infrared, optical, and ultraviolet regions of the spectrum. The major differences between the two classes are in the radio region and in x-rays, where radio-loud quasars tend to be brighter.

Radio emission generally comes from nonthermal sources. The Sun and other stars, as well as the coals in your BBQ grill, are examples of thermal emission — they radiate by virtue of the temperature of the emitting material. Nonthermal emission, on the other hand, is due to a different mechanism. Astronomers think the most likely mechanism for quasars' radio emission is synchrotron radiation created as electrons spiral around magnetic field lines. This radio emission apparently originates in the center of the quasar, close to the black hole. It is not reprocessed — absorbed, scattered, or re-emitted energy from the black hole or accretion disk — but rather a separate, primary energy source.

Recent observations at millimeter wavelengths, which span the gap between the radio and infrared, show that the radio and far-infrared emission in most

PHYSICAL PROCESSES

X-ray

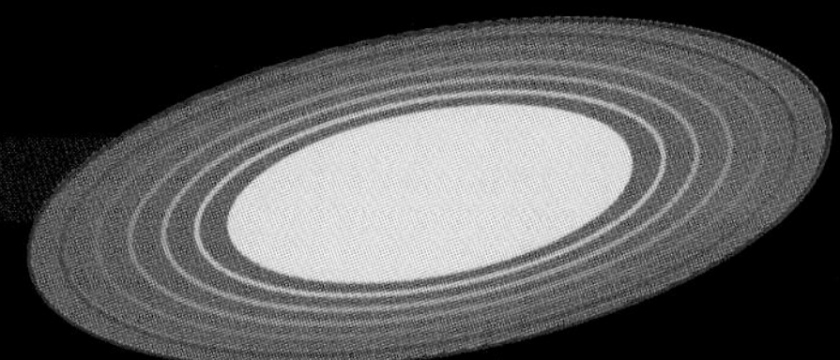

Many of the x-rays we see originate in the central source but then reflect off gas in the surrounding accretion disk.

Most ultraviolet radiation is thermal emission from the hot accretion disk surrounding the central black hole.

Ultraviolet

Much of the optical radiation in quasars comes from ordinary stars located in the host galaxy.

Optical

Ultraviolet

Infrared

Infrared

Infrared radiation emanates from dust in the host galaxy heated by high-energy radiation from the central source.

Radio radiation in radio-loud quasars derives from electrons spiraling around magnetic-field lines.

Radio

Theoretical models of the physical processes in quasars can account for the radiation we observe.

quasars arise from different mechanisms. Only in the subset of radio-loud quasars that have no extended radio emission (called core-dominated), does the same emission mechanism dominate both regions. This interpretation is consistent with the popular unified scheme for radio sources in which core-dominated quasars are believed to have extended radio emission that points in our direction and appears brighter as a result of its rapid motion toward us relative to the rest of the quasar. The infrared part of the nonthermal emission would be boosted similarly and thus dominate any other components that might contribute in this energy range. If this is true, the main infrared emission mechanism in core-dominated, radio-loud quasars is different from that of other quasars despite its similarity in brightness relative to the optical and ultraviolet emission.

Infrared Emission

If the infrared light in core-dominated, radio-loud quasars is dominated by nonthermal synchrotron emission, what causes the roughly equal amount of infrared light in other quasars? The infrared energy distributions frequently have a bumpy appearance rather than the smooth, straight shape expected for synchrotron emission. Because the energy distribution of thermal emission peaks at a frequency that corresponds to the temperature of the emitting material, an obvious way to get a bumpy distribution is from material emitting at a number of different temperatures. For temperatures corresponding to infrared emission, dust is the most likely candidate material. Dust is bound to be present in the host galaxy and is very likely to be heated by the strong central source. It absorbs this energy, warms up, and then re-emits the energy thermally. This is a form of reprocessed energy. A small amount of thermal emission from dust at several different temperatures combined with nonthermal emission can explain the bumpy shapes we observe.

However, new millimeter observations have raised the question of whether a nonthermal component contributes at all. Instead, perhaps all the infrared emission from quasars is thermal emission from dust. This pure dust model would explain the bumpy shape in the near-infrared as well as the sharp decrease in the energy distribution at wavelengths longer than about 100 microns shown by the new observations. A pure dust model would also explain the characteristic dip in the emission as you go from infrared to optical wavelengths (see the diagram on page 36). Dust is destroyed by sublimation at temperatures higher than about 2000 kelvins so there would be no dust to radiate at higher temperatures and give rise to optical light, which is presumably generated by a different process. This maximum temperature also predicts the position of the dip, which always occurs close to 1 micron.

Unfortunately, life is not as simple as this. There are also problems with a pure dust model for the infrared emission. Huge amounts of warm dust would be needed to generate all the energy seen in the infrared. Either the galactic disk must be warped so that large amounts of dust can be heated by the central source or else an additional dust component, such as a torus surrounding the central energy source, must be present. Although evidence for such a torus exists for a few Seyfert 1 galaxies, objects similar to quasars but with lower luminosities, none has been seen in quasars. Another problem with a pure dust model is that the infrared spectra of quasars do not show absorption features due to silicate grains, features that are observed in normal galaxies when dust is present. In order for the model to work then, the dust must have a very different composition from that normally found.

Another problem with the pure dust model is the observed correlation between the near-infrared and x-ray light emitted by quasars. In x-ray bright quasars, at least, this correlation is much stronger than that between the x-ray and emission at any other wavelength. Such a strong correlation implies that the same or related emission mechanisms dominate in the two spectral regions, which is impossible for pure dust emission. One obvious possibility for the source is nonthermal synchrotron emission similar to that in core-dominated, radio-loud quasars.

A Bump in the Light

The best evidence that a black hole is the central energy source of a quasar comes from the optical and ultraviolet parts of the spectrum, where quasars put out the greatest fraction of their energy. The curving shape and relative uniformity of this part of the

Thermal-dust
R O X
Radio Optical X-ray

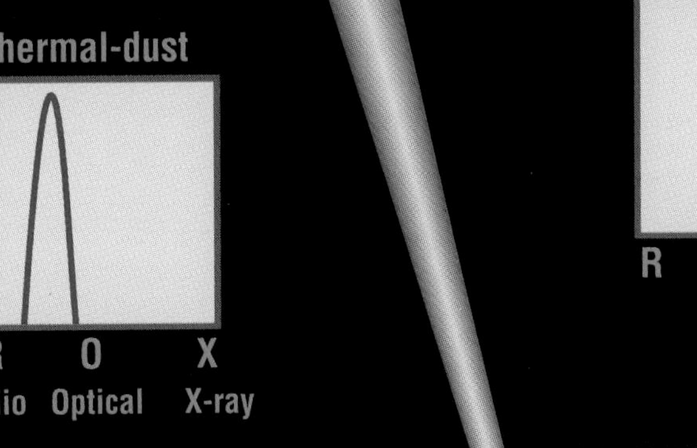

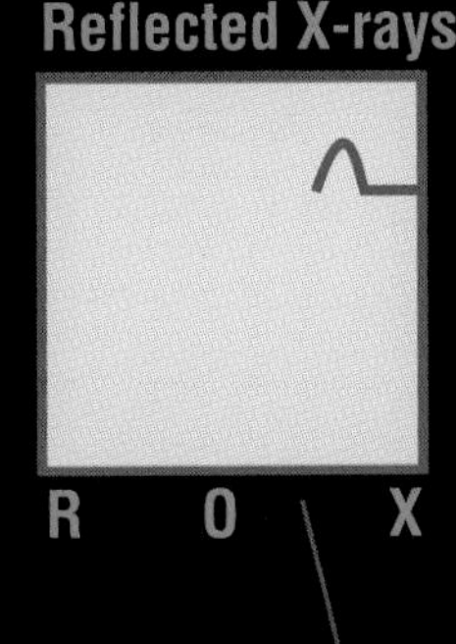

Reflected X-rays
R O X

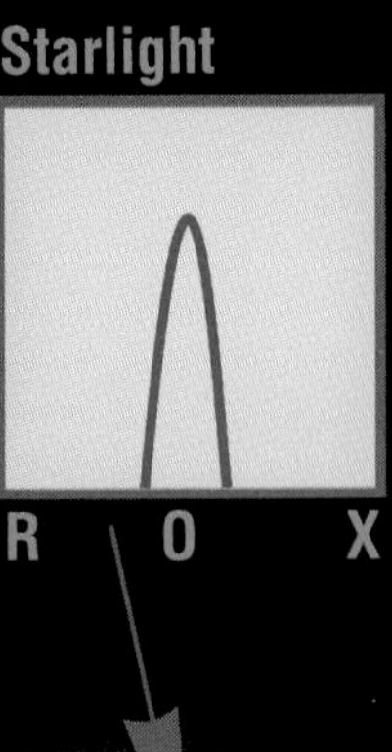

Starlight
R O X

Galaxy

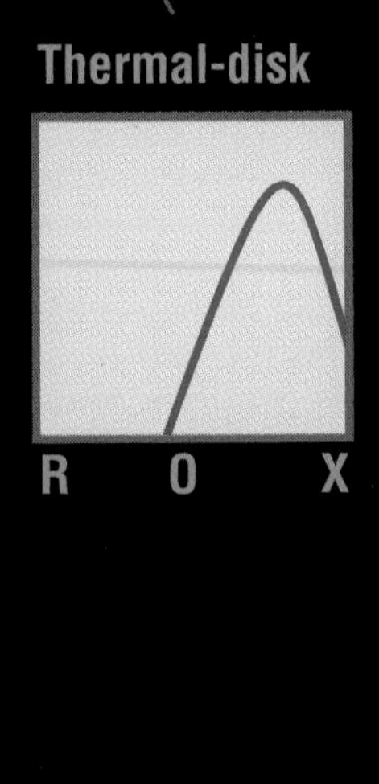

Thermal-disk
R O X

Synchrotron

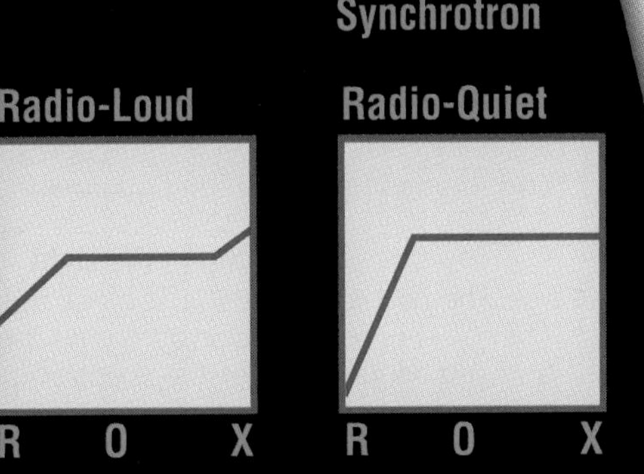

Radio-Loud
R O X
Radio-Quiet
R O X

The total output of a quasar is the sum of direct radiation from the central black hole, its accretion disk, and synchrotron radiation and re-emitted radiation from the surrounding dust, gas, and stars.

energy distribution compared with the irregular shape in the infrared led to the name "blue bump." The blue bump peaks somewhere in the extreme ultraviolet region and probably extends into the soft x-ray region around 40 angstroms. This curving, peaky shape suggests thermal emission, but the emission covers a wide range of frequencies and so must come from gas with a range of temperatures. This is very similar to that expected from an accretion disk surrounding a black hole. However, an accretion disk model cannot fully match the optical/ultraviolet distribution of a quasar unless it is combined with a nonthermal component. In this case a good, qualitative, match is obtained, though there are some discrepancies between the detailed predictions of current accretion disk models and observations.

An accretion disk also seems to be the best way to explain the x-ray emission in radio-quiet quasars. In this case, however, the accretion disk reprocesses energy from another primary energy source. Recent observations with the Japanese x-ray satellite Ginga show a strong emission line at 6.4 keV from iron and a broad bump at energies greater than about 10 keV. These features can be explained very nicely if x-rays emitted by a central source reflect off warm (about 10,000 kelvins), optically thick gas located in an accretion disk.

The energy absorbed in this reflection process heats the gas in the accretion disk, which then re-emits thermally in the extreme ultraviolet. This probably plays a role in the blue bump component because the physical conditions are similar. Thus the same accretion disk could be responsible for the blue bump, the soft x-ray excess, and the features in the x-ray emission spectrum. This model not only explains all the current x-ray data for radio-quiet quasars, but also ties in with the emission in the optical and ultraviolet regions. A major remaining question is what generates the original x-ray emission. Two possibilities are non-thermal synchrotron emission and electron/positron pair production, both of which could occur in a region close to and surrounding the black hole.

BURIED AT THE HEART of a quasar is a supermassive black hole that powers all its varied emissions.

Putting It All Together

Can we paint a picture of a quasar based on our current knowledge? Although the details still have to be ironed out, the simplified schematic diagram at left shows the likely ingredients and their contributions to the overall continuum emission of a quasar. These ingredients include as primary sources: thermal emission from an accretion disk surrounding a black hole that is heated both by the inward motion of the gas and perhaps by x-rays; non-thermal, synchrotron emission from a source in the core close to the black hole; and x-rays also from a source in the core close to the black hole. The significant sources of reprocessed energy are: reflected x-rays from the inner regions of the accretion disk and thermal emission from warm and cool dust present in the host galaxy and heated by energy from the central regions. Starlight from the stars in the host galaxy also contributes.

Qualitatively, this picture can explain all the data on quasar energy distributions now available to us. As more and better data become available over the next few years, we should be able to quantify the relative contributions and draw more detailed pictures of quasars themselves. In particular, observations of the soft x-ray, extreme ultraviolet, and ultraviolet regions from the ROSAT x-ray satellite and the Hubble Space Telescope already allow a more complete and detailed study of the blue bump. New infrared imaging instruments promise a vast increase in the amount of spectral data in that energy region with which to study the contribution of dust. New and more sensitive millimeter telescopes will also contribute here.

And this only touches on current capabilities. There are many new ground- and space-based facilities planned for the next decade that will help our study of the whole energy distribution. Given the progress over the past few years and the promise of the next few, this is certainly an exciting time in the field of quasar research. □

Belinda J. Wilkes is an astrophysicist at the Harvard-Smithsonian Center for Astrophysics in Cambridge, Massachusetts.

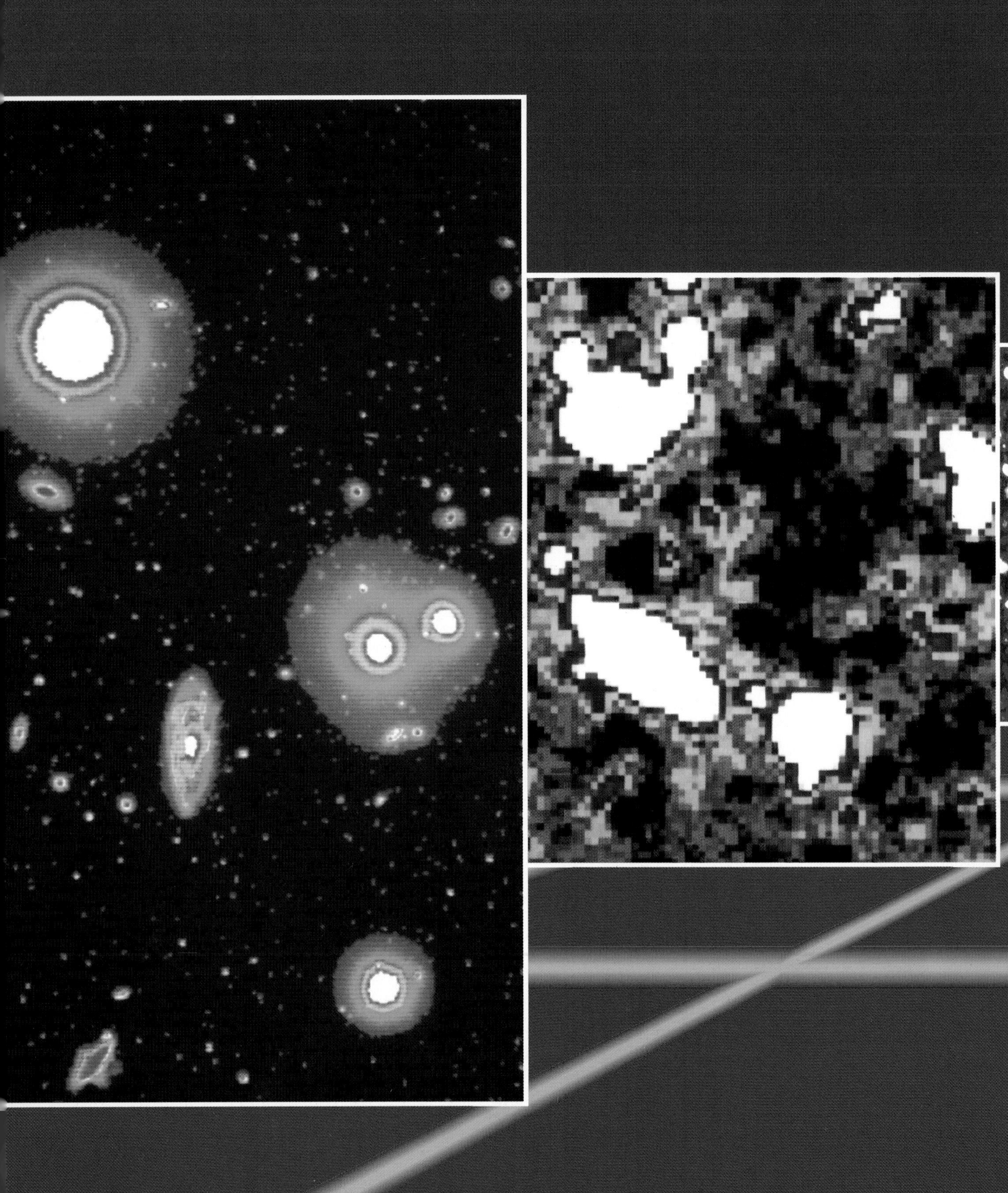

Beyond the Big Bang

New observations cast doubt on conventional theories of how the universe formed.

By Jeff Kanipe

BEYOND THE VISIBLE HORIZON of the universe, beyond the most distant galaxies and quasars we can see, lies the smooth and ancient background glow from the Big Bang. Astronomers want to know how the patterns of galaxies formed out of that smooth texture of energy and matter. Illustration by Steve Davis.

"It seems to be gratuitously courting disaster to expose our theories to conditions in which any slight weakness is likely to become magnified without limit. But that is just the principle of testing."

— *Sir Arthur Eddington,* The Expanding Universe.

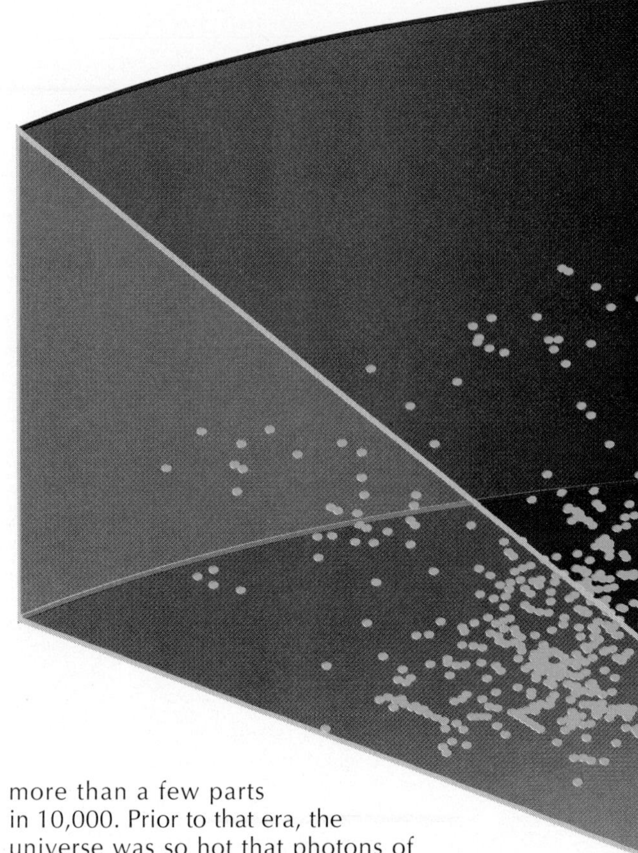

On November 18, 1989, the Cosmic Background Explorer (COBE) was launched into Earth orbit. Its mission: to study the dominant form of radiation in the universe, the cosmic microwave background that is believed to be the remnant heat from the Big Bang itself.

In just nine minutes of observations, COBE determined that the cosmic background radiates at a temperature of just a little over two and a half degrees above absolute zero. More importantly, it showed that the intensity of the background across the spectrum precisely matches that of a perfect emitter and absorber of radiation — an idealized object physicists call a "blackbody."

The COBE results were a stunning victory for the standard Big Bang model of the universe. Here was evidence that the primordial fireball was a uniform explosion of matter and energy. Out of a quantum seed the universe unfolded with the grace of a tropical blossom, balanced and smooth, but with all the structural attributes of a seaside fog. For astronomers, however, the COBE results added to a growing paradox in cosmology.

Paradox Found

The problem stems from COBE's confirmation that the background radiation is coming at us with equal intensity everywhere we look. This demonstrates that the early universe must have been smooth — matter and energy were once evenly distributed.

This is hardly the case today. You need only walk out into your backyard at night and explore the sky with a telescope to see that for yourself. The local universe is not smooth. Neither is the more distant universe we can see with giant telescopes. Lumps of matter exist in the visible universe — they're called galaxies and galaxy clusters.

Astronomers have mapped thousands of galaxies in three-dimensional space from both the Northern and Southern Hemispheres and have discovered that on large scales the universe is a vast, frothy tangle of galaxy clusters interspersed with voids of seemingly empty space.

As we peer farther into space and further back in time, we still find galaxies. One to two billion years after the Big Bang, astronomers see galaxies shining with the blue light of hot young stars. Even further back, shining like airport beacons from the very edge of the visible universe, astronomers still see lumps of matter — quasars.

The COBE observations, however, show that 300,000 years after the Big Bang, matter and radiation were distributed with variations in intensity of no more than a few parts in 10,000. Prior to that era, the universe was so hot that photons of energy were continually absorbed and re-emitted by matter, preventing stable atoms from forming. No lumps of matter could survive in the turmoil of the early universe. Hence, galaxies must have condensed *after* this epoch, when matter and radiation went their separate ways.

Thus the paradox: Galaxies form in clumps. Yet we know from COBE that the universe started out smooth and featureless. At the present time, astronomers cannot explain how the universe got from one state to the other because there doesn't appear to be enough time between the COBE era and the quasar/galaxy era for gravity to have collected matter into the complex clusters seen today. It is a problem serious enough to prompt a few researchers to suggest there is something wrong with the Big Bang theory.

The Three Pillars of Cosmology

Despite this so-called "structure problem," most astronomers are reluctant to abandon the Big Bang. The evidence supporting the theory, they say, is extensive and compelling enough to stand on its own. Astronomers cite three observations as the underpinnings of the Big Bang cosmology.

The first was made in the 1920s, when astronomers noted that the spectral lines of most galaxies (the lines arising from the emission or absorption of radiation by certain atoms at fixed wavelengths) appeared shifted toward the red end of the spectrum, a shift attributed to the galaxies apparently speeding away from us.

Then, in 1929, American astronomer Edwin Hubble discovered that a galaxy's recession velocity is directly proportional to its distance. Astronomers concluded that the universe is expanding and that

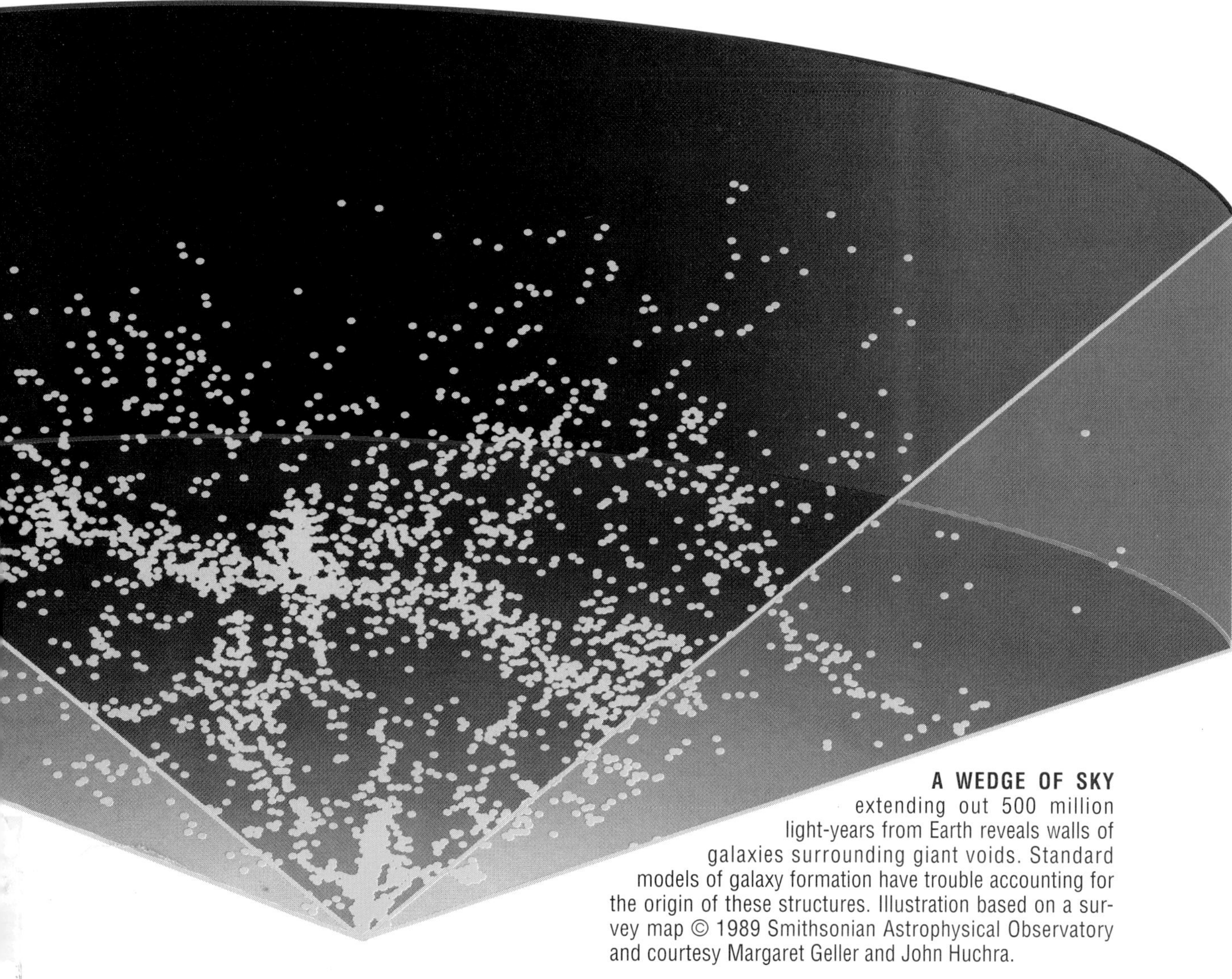

A WEDGE OF SKY extending out 500 million light-years from Earth reveals walls of galaxies surrounding giant voids. Standard models of galaxy formation have trouble accounting for the origin of these structures. Illustration based on a survey map © 1989 Smithsonian Astrophysical Observatory and courtesy Margaret Geller and John Huchra.

space between the galaxies is steadily increasing like the space between raisins in a cinnamon roll as it rises. For the universe to be expanding, however, a propulsive force of unimaginable magnitude — a Big Bang — must have set matter on its runaway course.

The second fundamental observation supporting a Big Bang is the measurement of the total abundance of light elements in the universe. This is akin to analyzing the ingredients of a cake after it's been baked. You know that for the cake to have the consistency, flavor, and structure that it has, certain substances must be present. From an analysis of samples of the cake you can derive a ratio of one ingredient to another, which can then be applied to the entire body of the cake.

Cosmologically speaking, the elements present in the early universe can be sampled by looking at the most ancient components of the universe — the oldest stars (usually found in globular clusters). The ratio of helium to hydrogen in these stars is exactly what the Big Bang theory predicts should have been forged by thermonuclear reactions 100 seconds after the primordial event. Though scientists have attempted to come up with other theories to account for the light-element abundances, none has accounted for them as well as the Big Bang.

Finally, we return to the cosmic background radiation, considered the most compelling argument for a Big Bang. The cosmic background temperature is 2.7 kelvins, and this radiation is uniform across the entire sky. Nowhere is the temperature higher or lower than this value. Slight variations would indicate hot spots in the background, the lumpy seedlings for the first clusters of galaxies. So far, none have been detected. This is exactly what the Big Bang model predicts.

Cracks in the Pillars?

But it is here that problems arise. For despite the three observational pillars, the Big Bang theory fails to fulfill the requirement for a complete cosmology: it does not explain it *all*. Everything in existence. Everywhere. The Big Bang theory explains only the first few moments of the universe; it stops short of explaining how galaxies and stars materialized in the first one billion years of the universe. This leads to the unsettling conclusion that, though astronomers believe they know how the universe began, they don't know how it got to its present state. Like the missing 18 minutes from the Watergate tapes, there is a narrow but crucial gap in the transcripts of the universe.

Filling the gap requires a theory that links the Big

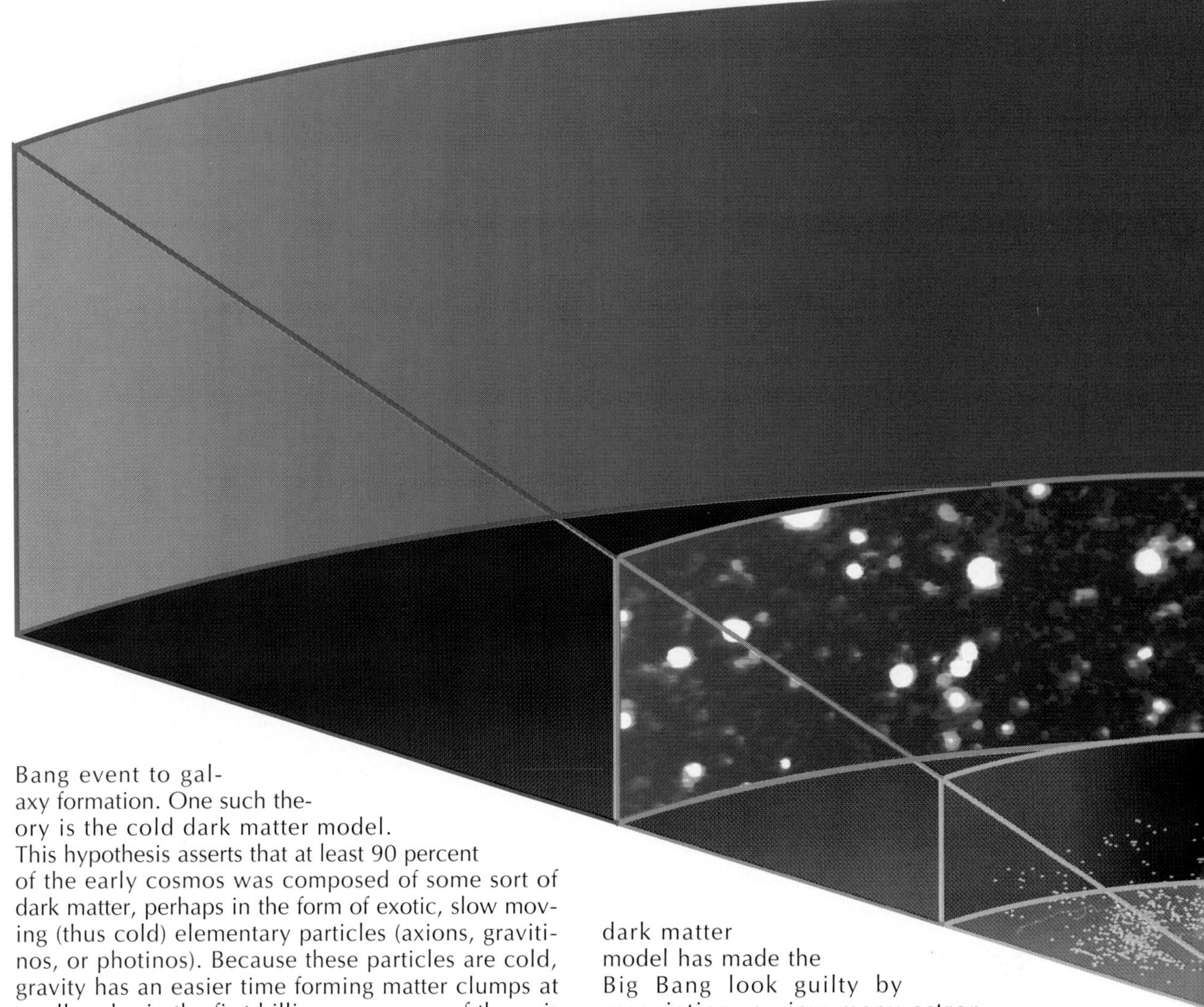

Bang event to galaxy formation. One such theory is the cold dark matter model. This hypothesis asserts that at least 90 percent of the early cosmos was composed of some sort of dark matter, perhaps in the form of exotic, slow moving (thus cold) elementary particles (axions, gravitinos, or photinos). Because these particles are cold, gravity has an easier time forming matter clumps at small scales in the first billion years or so of the universe. These clumps — estimated to have the mass of a typical globular cluster — gravitationally attract more matter until galaxy-sized masses are built up. The galaxies themselves then collect into the elaborate clusters we see today. This is known as the "bottom-up" model of galaxy formation.

Unfortunately, cold dark matter, exotic or otherwise, has yet to be detected, save indirectly by what are thought to be its gravitational effects on normal matter, specifically the motions of galaxies in clusters and the rotations of spiral galaxies. Astronomers have yet to determine what form this dark matter takes. (See "Shedding Light on Dark Matter," February 1992 ASTRONOMY.)

Compounding the problem, even the most generous cold dark matter models cannot account for the astonishing dimensions of large-scale structures that are turning up in galaxy surveys today — structures that span nearly a billion light-years.

Is the Big Bang in trouble? Most astronomers think not. They view the Big Bang and galaxy formation as separate events. "I don't think anything that's happened recently has changed the majority opinion that the Big Bang is sound," says Ethan Vishniac of the University of Texas at Austin. "I would say the current situation is that we have no believable theory of galaxy formation. What has happened in the last year or so is that . . . the cold dark matter theory has collapsed."

For some astronomers, the downfall of the cold dark matter model has made the Big Bang look guilty by association, a view many astronomers reject. "Cold dark matter is a theory about the beginning of the formation of structure," says Princeton University cosmologist James Peebles. "The Big Bang is something quite different. The problem is that the press has conflated the two. They take the discussions about the beginnings of structure and confuse them with the beginning of the universe."

Taking issue with Vishniac, Peebles, and practically the entire astronomical community is Anthony L. Peratt, a physicist and cosmologist at Los Alamos National Laboratory. Rather than propose a better theory of galaxy formation, Peratt suggests scrapping the Big Bang altogether.

"The Big Bang theorists attempt to decouple themselves from the problem of galaxy formation," says Peratt. "But one must ask what kind of cosmology is it that cannot account for the galaxies and stars we observe? This was a basic requirement for all previous cosmologies, from the time of the Ionians to Newton. Big bang cosmologists have struggled mightily to find a mechanism for galaxy formation in spite of their arguments that this is not their problem."

Peratt belongs to a group of scientists who subscribe to an alternative cosmogenesis theory called "plasma cosmology," which argues that the laws of electromagnetism, rather than gravity, dominate and shape the universe.

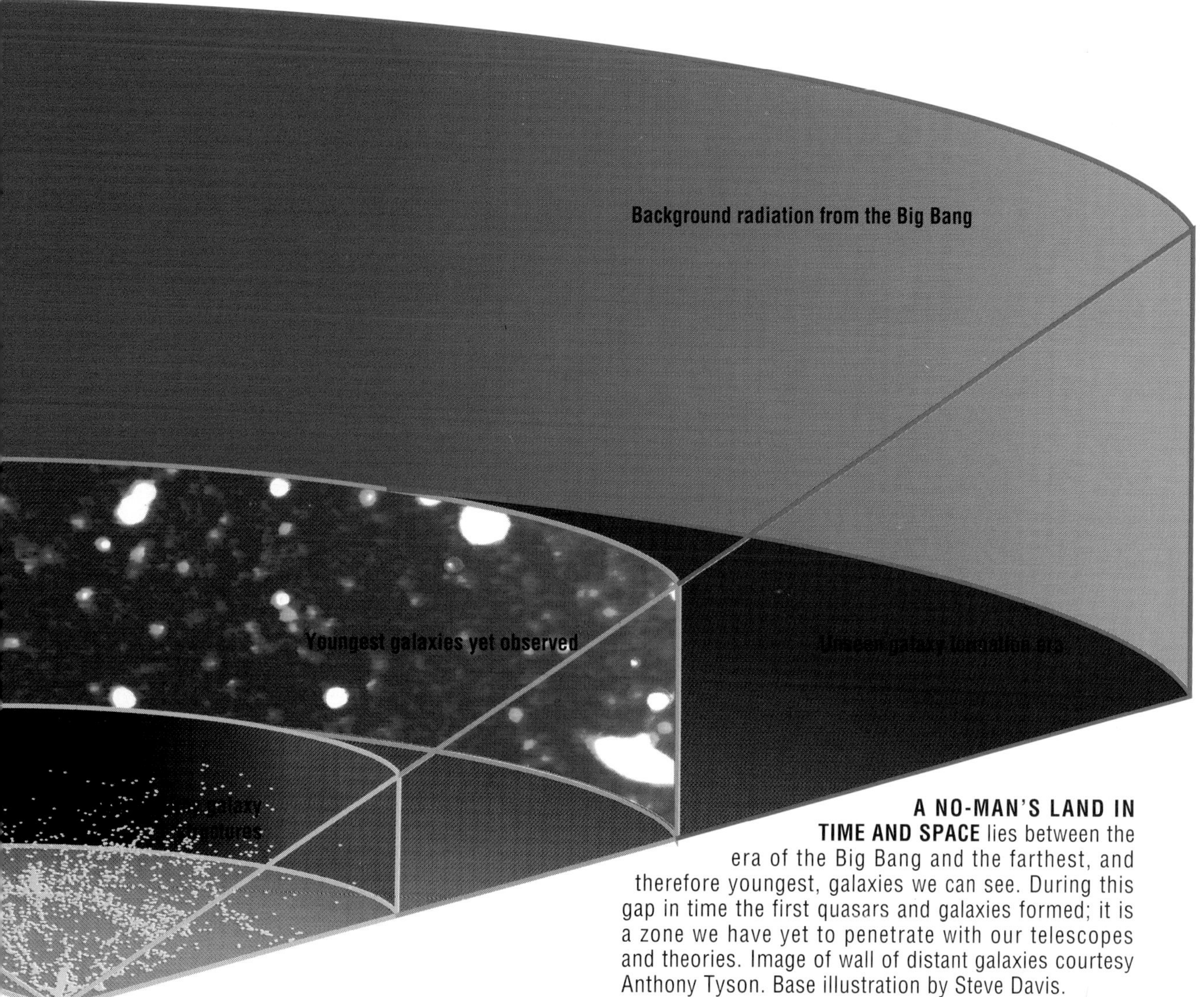

A NO-MAN'S LAND IN TIME AND SPACE lies between the era of the Big Bang and the farthest, and therefore youngest, galaxies we can see. During this gap in time the first quasars and galaxies formed; it is a zone we have yet to penetrate with our telescopes and theories. Image of wall of distant galaxies courtesy Anthony Tyson. Base illustration by Steve Davis.

An Alternate Universe

Plasma is a state of matter that resembles a gas, except that instead of consisting of electrically neutral atoms, plasma contains charged particles — electrons torn from atoms in a gas, leaving a cloud of negatively charged free electrons and positively charged ions. Although affected by gravity, these charged particles can also be accelerated by electromagnetic fields.

Most of the visible mass of the universe is plasma. On Earth, the most dramatic forms are lightning and aurorae. In space, stars are gravitationally bound plasmas. On larger scales, plasmas have been detected at the galactic center and in the double lobes of radio galaxies. The largest plasma structure was discovered in 1989, when faint radio emission was detected between two superclusters of galaxies, indicating that the superclusters are embedded in a warm plasma.

The plasma cosmological model was first proposed by Hannes Alfvén, of the Royal Institute of Technology in Stockholm, Sweden. In 1970, Alfvén was awarded the Nobel Prize in physics for his work in solar magnetohydrodynamics — the movement of plasma in magnetic fields. In separate work, Alfvén reasoned that sheets of electric currents crisscross the universe. These electromagnetic fields and not gravity, plasma cosmologists suggest, are the major movers of matter on the grandest scales. Interaction with electromagnetic fields, Peratt says, enables plasmas to exhibit complex structure and motion that far exceed what's possible with gravity alone.

Plasma proponents have conducted computer simulations involving tens of millions of test particles (used to represent galaxies) to demonstrate that a fundamentally uniform plasma will eventually break up into cellular structures that look remarkably similar to the large-scale structures seen today.

The cellular structures would, in turn, act to scatter radiation. Plasma cosmologists theorize that what we detect as the cosmic background actually results when local fields and currents scatter microwave radiation from pervasive plasmas like a dense fog scatters a car's headlights.

But what about the successful prediction of the abundance of light elements proclaimed by Big Bang proponents? "We have always questioned just how well known the abundances of light elements are in the universe," says Peratt. "And if the universe is cellular and filamentary, what are the abundances of the elements within and without the filaments?" Peratt cites a recent measurement of light elements in galaxies that indicates a helium abundance *lower* than that allowed by the standard Big Bang model. A lower assumed density of the universe would produce such a lower helium abundance, but that would also mean the standard Big Bang model does not satisfy the primordial abundance constraints for the other light elements for any density. Interestingly, such low helium abundances were found by other researchers in 1988.

As for the "missing link" between the Big Bang and the first galaxies, plasma physicists claim there is no missing link because there never was a matter-forming epoch. As Peratt writes in an article in the January/February 1990 issue of *The Sciences*, "There is no expansion, and there need not be any final crunch. Unlike the universe envisioned in the Big Bang model, the plasma universe evolves without beginning and without end: it is indefinitely ancient and has an indefinite lifetime in store."

Peratt's plasma cosmology isn't the only alternative idea. The steady state model still has a few supporters. Steady state theorists claim that new galaxies are constantly forming to replace old galaxies at a rate determined by the expansion rate of the universe. Supporters also argue that the cosmic microwave background is actually normal radiation from stars, radiation that has been scattered by metallic "whiskers" sown throughout the interstellar medium in supernovae ejecta.

But don't confuse plasma cosmology with steady state theories, explains Peratt. Steady state invokes a perfect cosmological principle in which the universe ap-pears the same at all points at all times — past, present, and future. In contrast, the plasma universe does evolve with time. According to Peratt, "The plasma universe model has nothing to do with the steady state model. Both the Big Bang and the steady state espouse gravity as the sculptor of the universe."

Paradox Lost?

Among the supporters of alternative cosmologies, there are few agreements except, perhaps, in their enthusiasm for refuting conventional cosmology. Each has a response to the arguments posed by Big Bang cosmologists.

Mainstream astronomers counter these attacks by pointing out that alternative theories like the plasma and steady state cosmologies don't come close to the predictive power of the Big Bang. It has withstood three-quarters of a century of scrutiny, they say. It will take more than a few stabs to topple it.

"We don't really have a complete theory of planet formation but we wouldn't use that to question that the Sun formed from contracting interstellar gas," says Lennox Cowie of the University of Hawaii. Cowie has made a specialty of probing the distant universe for primeval galaxies. Recently, he and a team of astronomers completed a deep infrared imaging survey that turned up thousands of faint dwarf galaxies approximately one hundredth the size of our Galaxy lying at moderate distances. These faint galaxies are likely the most populous type of galaxy in the universe. But more importantly, he adds, "They probably contain at least as much

FROM SMOOTH TO FOAMY, a series of computer simulations shows the transition from the uniform structure that arose out of the Big Bang (bottom left) to the filaments and voids of the universe today (top right). Current models of this process are now under suspicion. Images of "cold dark matter" simulation courtesy James Gelb, MIT.

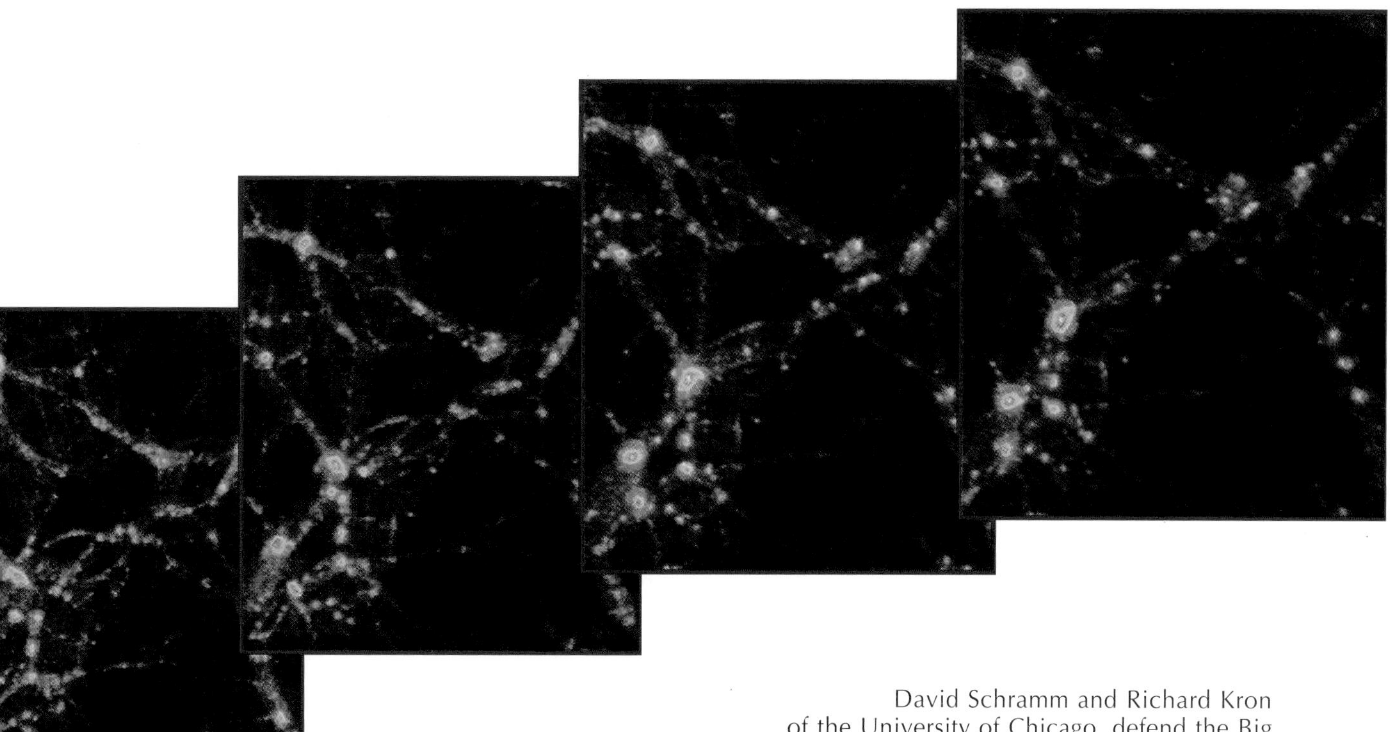

mass as normal galaxies." A universe filled with dwarf galaxies may account for a great deal of the missing mass in the universe, mass predicted by the cold dark matter model. And if you have more mass, you can account for more structure. So perhaps cold dark matter is still lurking out there somewhere, waiting to be found.

The Big Bang theory could also be strengthened by looking at galaxy formation in a new way. In December 1991 a team of radio astronomers at the Very Large Array in New Mexico announced the discovery of a large mass of hydrogen gas near the edge of the observable universe. According to the "top-down" theory of galaxy formation, galaxy clusters formed when giant primordial gas clouds broke apart into smaller gas clouds and not when smaller clouds came together as in the "bottom-up" scenario. A refined top-down idea may explain how large structures can form in a relatively short time span, solving the current problem. The new discovery may be the first evidence for the process.

The bottom line is that the majority of astronomers are reluctant to toss out the Big Bang theory because of an inability to explain how galaxies form. "The problem of galaxy formation is one that depends on rather poorly understood physics and also on some rather shaky assumptions," Cowie says. "The Big Bang theory on the other hand predicts both the light element abundances and the microwave background. I think until we get a much better understanding of the history of galaxy formation it would be very presumptuous to question the underlying cosmological model."

Princeton University's James Peebles agrees. In fact, in a paper published in the August 29, 1991 issue of the British journal *Nature*, Peebles, along with fellow Princeton cosmologist Edwin Turner, and David Schramm and Richard Kron of the University of Chicago, defend the Big Bang. They point out that plasma cosmology can't explain why the cosmic background radiation is so uniform over the entire sky. And that is just one of many problems they see with the alternative cosmologies.

As Peebles puts it, "Time and again people have said, It's difficult to think of a theory of galaxy formation that can be consistent with all observations. So perhaps that difficulty means that the framework within which we're working — the Big Bang — is wrong.' And time and again that inference has been shown to be incorrect.

"By being a little more clever in the invention of theories," Peebles says, "a person can find a way around the problems and continue to look for a way to build within the Big Bang framework."

Peratt, however, remains steadfast. "Plasma cosmologists don't feel we are 'closing in on creation. . . .' In fact, if our findings are anywhere near right, the universe is much bigger and older than previously imagined."

The conflict between the differing views of the universe presents a classic example of an established theory undergoing a revolution. Though the Big Bang model may have endured for 75 years, history is replete with erroneous scientific beliefs that survived centuries, as well as some "crackpot" theories that later became accepted (continental drift, for example).

Is the Big Bang theory wrong? It is certainly under attack. Scientists are probing the Achilles' heel of this venerable cosmology in a debate sparked by what new telescopes probing to the edge of the universe have shown us.

But debate is what science thrives on. Without it, we would be like a people starving to death in a land of sacred cows. □

Jeff Kanipe, a former associate editor of ASTRONOMY, is now the editor of Star Date *magazine, published by McDonald Observatory and the University of Texas at Austin.*

Too Smooth
COBE's Perfect Universe

If the early universe was as uniform as recent observations from the Cosmic Background Explorer indicate it was, where did the galaxies come from?
by Jeff Kanipe

HOW THE UNIVERSE LOOKS depends on the wavelength of the "window" you look through. Optical telescopes, like the *Hubble Space Telescope* (opposite page), see varied structure in all directions. *COBE*'s microwave instruments (below), however, see only a smooth, featureless universe — the diluted glow of the Big Bang. Painting by Paul DiMare.

NASA

The COBE Results

COBE's sky maps of the background radiation indicate the extraordinary smoothness of the early universe.

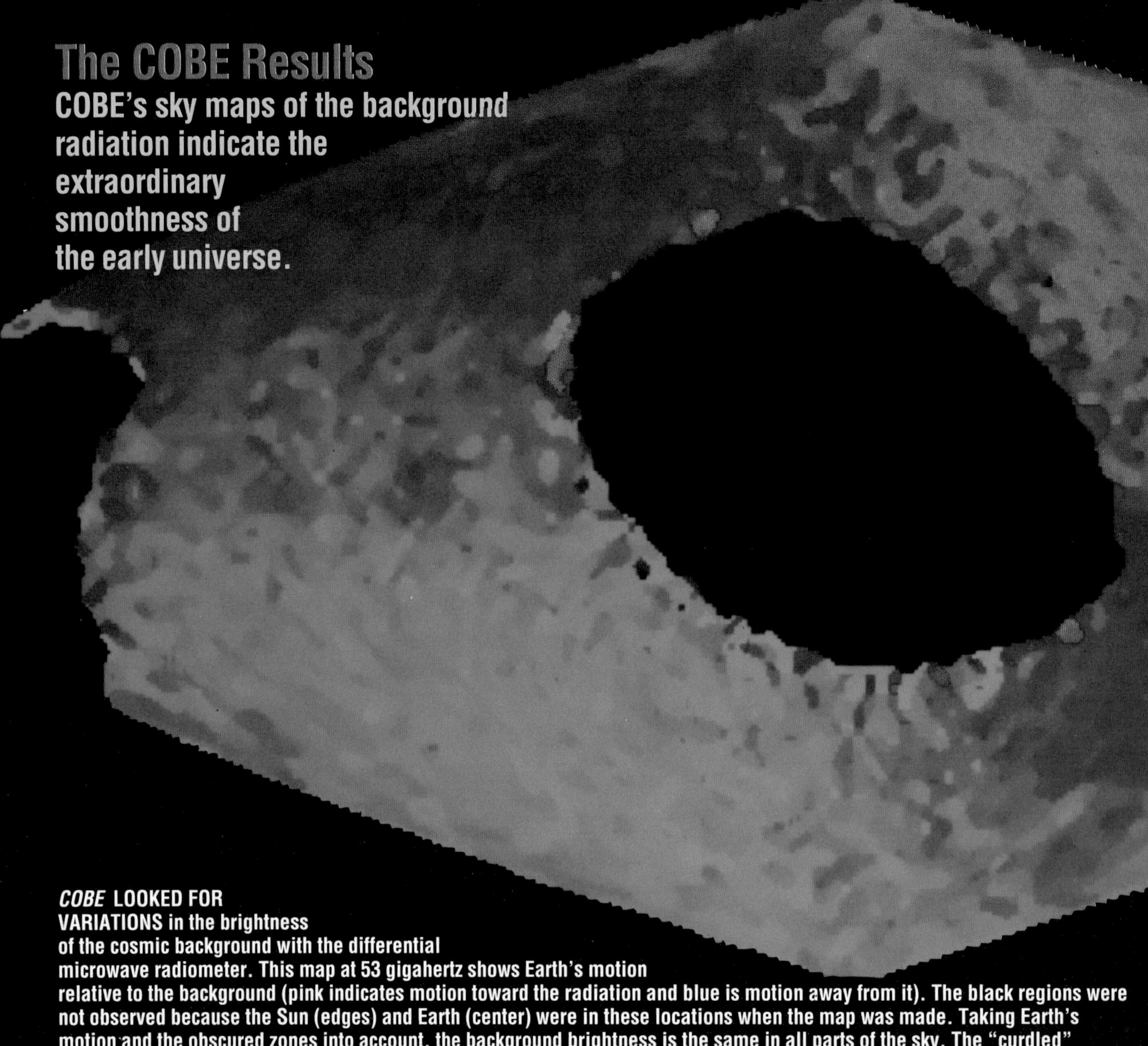

***COBE* LOOKED FOR VARIATIONS** in the brightness of the cosmic background with the differential microwave radiometer. This map at 53 gigahertz shows Earth's motion relative to the background (pink indicates motion toward the radiation and blue is motion away from it). The black regions were not observed because the Sun (edges) and Earth (center) were in these locations when the map was made. Taking Earth's motion and the obscured zones into account, the background brightness is the same in all parts of the sky. The "curdled" features are areas of sky below *COBE*'s angular resolution that have not yet been fully mapped. NASA photo.

When the *Cosmic Background Explorer* first looked into deep space last December, it saw a perfect universe, one without features to give it shape or variation to give it diversity. To *COBE*'s eyes the afterglow of the Big Bang was a vast, smooth backdrop radiating everywhere at a temperature barely above absolute zero. The definitive measurements of this background radiation finally confirm what astronomers have long suspected about the universe. Many are already ranking *COBE*'s findings as among the most important cosmological achievements of the century.

Since its launch on November 18, 1989, *COBE* has exceeded all expectations. The radiation from the Big Bang has had 15 billion years to expand and cool and is today remarkably feeble — about 100 million times fainter than the heat produced by a typical birthday candle. Nevertheless, in only nine minutes of sky observations *COBE* nailed down the microwave background temperature at 2.735 kelvins, with an uncertainty of only 0.06 kelvins. Moreover, *COBE* confirmed that this temperature is remarkably uniform, or isothermal, across the entire sky.

The task now facing astronomers in the wake of the *COBE* results is to explain how galaxies could have formed in such a serene universe. According to current galaxy formation theories, the smoothness of the early universe must have been stirred up in some way, perhaps by explosions of supermassive objects. These disturbances would have served as the seeds around which matter collapsed into galaxies. The energy released by these initial disturbances should be detectable today in the cosmic microwave background, but *COBE* didn't find any. Clearly, cosmologists are going to have to come up with a more elaborate way to make their galaxy formation theories fit the *COBE* observations.

What *COBE* Saw

Without a doubt the most spectacular of the *COBE* results was the isothermal background spectrum, which was generated from data taken by *COBE*'s far infrared absolute spectrophotometer (FIRAS). According to John Mather, *COBE* project scientist and FIRAS principal investigator, the instrument "can look back and measure the intensity of the radiation one year after the Big Bang," which is farther back than any other instrument has ever measured.

Comprised of sixty-seven measurements across twenty different frequencies, the spectrum shows that

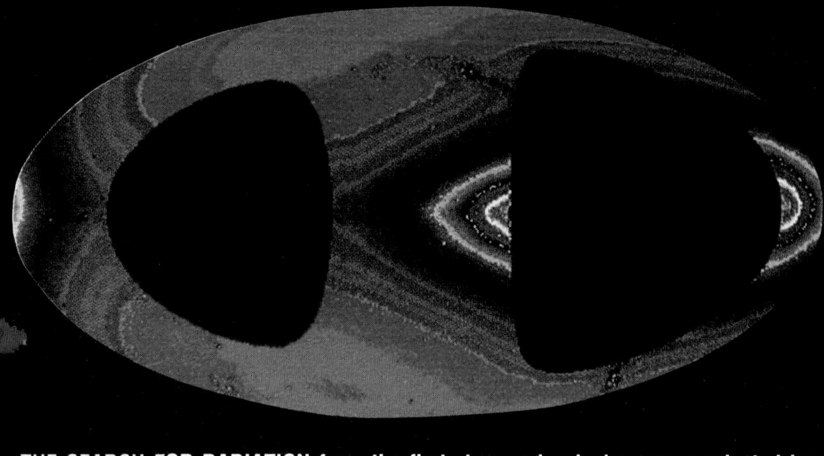

THE SEARCH FOR RADIATION from the first stars and galaxies was conducted by the diffuse infrared background experiment aboard *COBE*. The squatty, cone-shaped contour extending right to left through the center of the image is emission from interplanetary dust grains in the ecliptic plane. Emission from interstellar dust in the Milky Way is prominent in the ring-shaped feature, which is interrupted by the position of the Sun and Earth (right and left blank areas, respectively). The Milky Way in Cygnus is at top, and Carina is at bottom. The Large Magellanic Cloud is the small bright patch at bottom center. The "spots" throughout the map are individual stars in our Galaxy. Looking beyond these local structures, no significant features are seen. NASA photo.

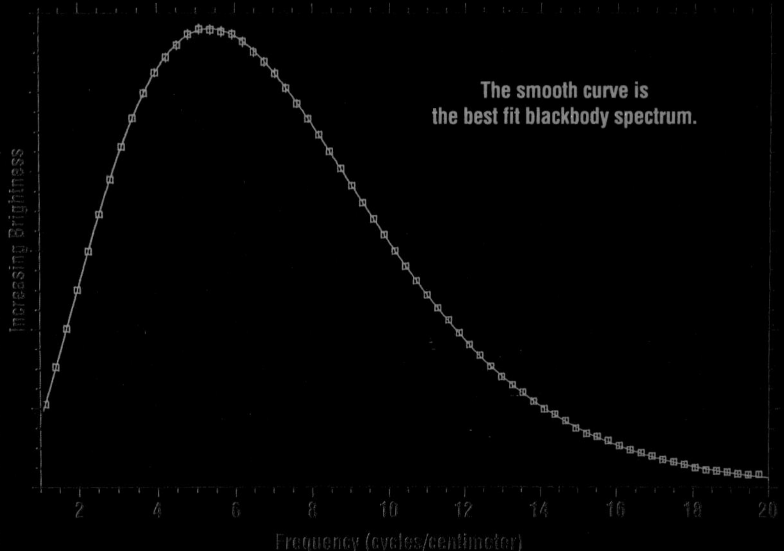

COBE FOUND A PERFECT BIG BANG. Sixty-seven measurements (the small squares) by the far infrared absolute spectrophotometer show that the temperature of the cosmic background does not deviate from a blackbody spectrum (smooth line) by more than 1 percent. This indicates a smooth, uniform Big Bang. Lumps in the distribution of matter in the early universe would have produced significant variations from the blackbody spectrum. Diagram courtesy of John Mather.

in the wavelength range between 0.5 millimeters and 5 millimeters the intensity of the background radiation varies less than 1 percent from a "blackbody" spectrum. A blackbody is an idealized opaque object that continuously absorbs all the radiation that strikes it (that's why it's called black) and then reradiates all the energy, leaving neither a net gain nor net loss of energy in the object.

The fact that the intensity of the microwave background precisely matches that of a perfect absorber and emitter attests to the fact that matter and radiation in the early universe were in equilibrium, i.e., both had identical temperatures. Thus, according to the FIRAS measurements, the Bi[illegible]g was a smooth and unifor[illegible]t. Had deviations from the blackbody spectrum been detected, these would have indicated energetic processes in the early universe and a clumpier distribution of primordial matter. But no such deviations were detected.

Another instrument aboard *COBE,* the differential microwave radiometer (DMR), was designed to detect variations in the brightness of the cosmic background radiation. George Smoot, principal investigator for the instrument, reported that maps taken at 31, 53, and 90 gigahertz show that the background glow is equally bright in all directions. The previously known "di-[illegible] (or 180-degree) variation was detected, but this is simply a Doppler effect caused by the motion of Earth relative to the background ra-

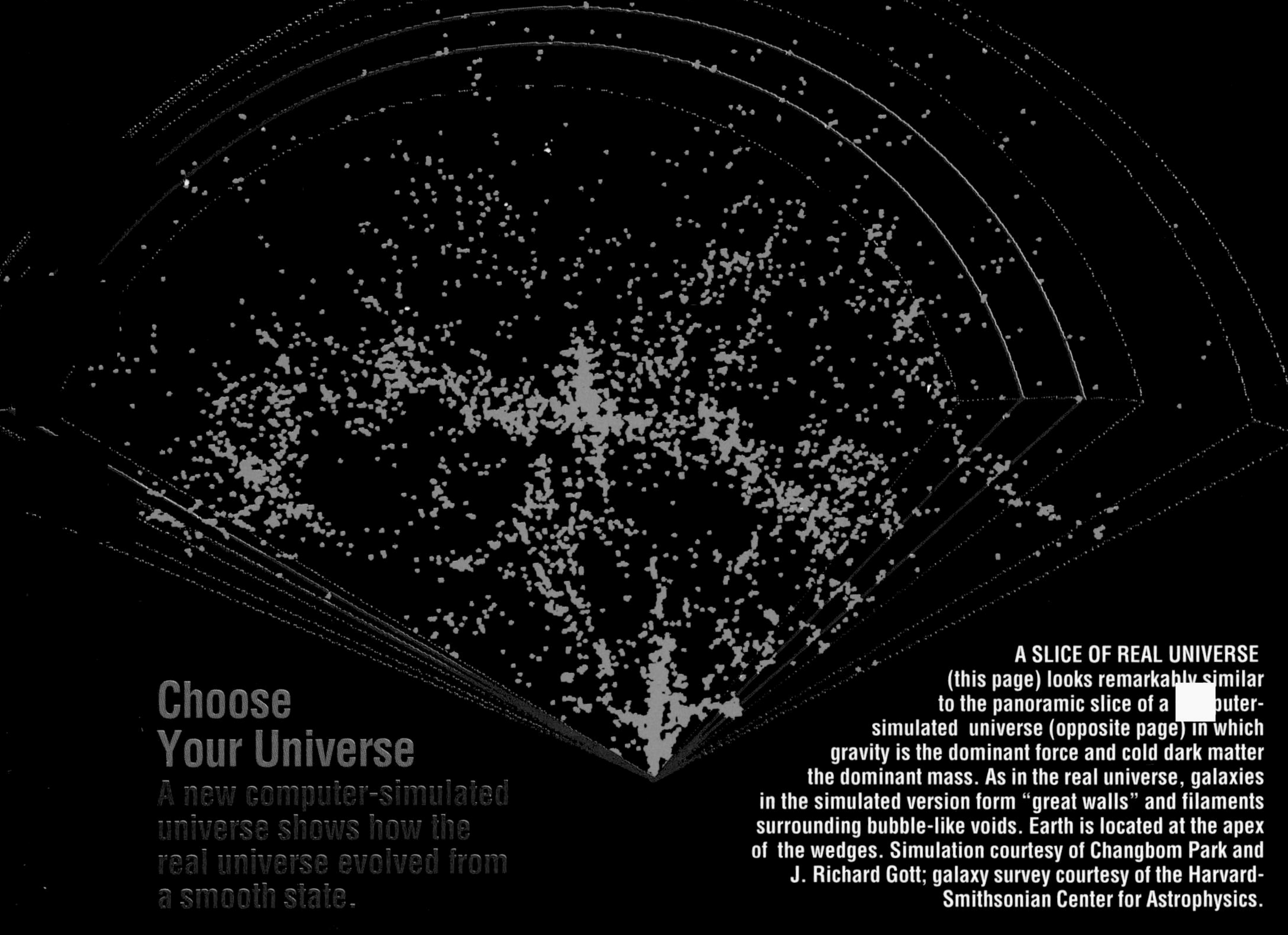

Choose Your Universe

A new computer-simulated universe shows how the real universe evolved from a smooth state.

A SLICE OF REAL UNIVERSE (this page) looks remarkably similar to the panoramic slice of a computer-simulated universe (opposite page) in which gravity is the dominant force and cold dark matter the dominant mass. As in the real universe, galaxies in the simulated version form "great walls" and filaments surrounding bubble-like voids. Earth is located at the apex of the wedges. Simulation courtesy of Changbom Park and J. Richard Gott; galaxy survey courtesy of the Harvard-Smithsonian Center for Astrophysics.

diation. Subtract that and the radiation is uniform across the sky.

A third instrument, the diffuse infrared background experiment (DIRBE), looked for infrared radiation from the luminous primeval stars and galaxies thought to have formed in the early universe. The instrument's principal investigator, Michael Hauser, said that maps of the sky at 1.2, 12, and 240 microns clearly revealed bright radiation from solar system dust, stars, and interstellar dust. But again, no background structures were apparent.

All You Need Is Gravity

COBE's findings agree with the cosmological principle, which asserts that there are no preferred vantage points in the universe: The view of the cosmos is the same in every direction no matter where the observer is located. Yet when we peer into different parts of the sky with telescopes, we see a variety of structure. Bright stars and clumps of dust and gas favor the disk of the Milky Way, while in directions perpendicular to the disk we see "great walls" of galaxies shot through with immense volumes of empty space. These structures, in fact, may be more extensive and complex than astronomers believed. During the same scientific meeting at which the *COBE* results were announced, an international team of astronomers reported their discovery of *many* walls of galaxies interspersed through billions of light-years of space. (See AstroNews, page 10.)

How, then, can the uniformity of the cosmic microwave background be reconciled with the existence of planets, stars, and galaxies? For these objects to exist, there must have been density ripples or wrinkles in the early universe on some scale. What force could have disturbed the universe enough to spawn the rich structure we see today and yet go undetected?

One theory, called explosive amplification, is that colossal blast waves in the early universe swept up and compressed matter into galaxies. Astronomers don't have to look any farther than our own Galaxy to see evidence for this type of explosion on stellar scales. Could a series of primordial explosions of a few exceptionally supermassive objects have ripped through the uniformity of the early universe and formed galaxies on the crests of the resulting shockwaves?

Although such explosions would certainly have had the power to create large-scale structures, the *COBE* results clearly rule them out. Explosions of primeval quasars or supermassive stars would have been very large-scale events that probably would have been seen by the diffuse infrared background experiment. Moreover, the explosion scenario begs the question of how the objects that exploded formed. Unless you resort to further theoretical convolutions, you're back to square one.

Many astronomers, however, are taking a conservative approach. Gravity, they say, can do the trick without all the mayhem. For astronomers like J. Richard Gott of Princeton University, *COBE* confirmed what they had suspected all along. Gott went so far as to say that *COBE's* results are exactly what you would expect to find if gravity were the primary force moving matter around in the early universe.

"Gravity moves stuff around quietly without redistributing the spectrum of the thermal radiation," said Gott, referring to the isothermal spectrum observed by *COBE.* "If there were violent bursts of en-

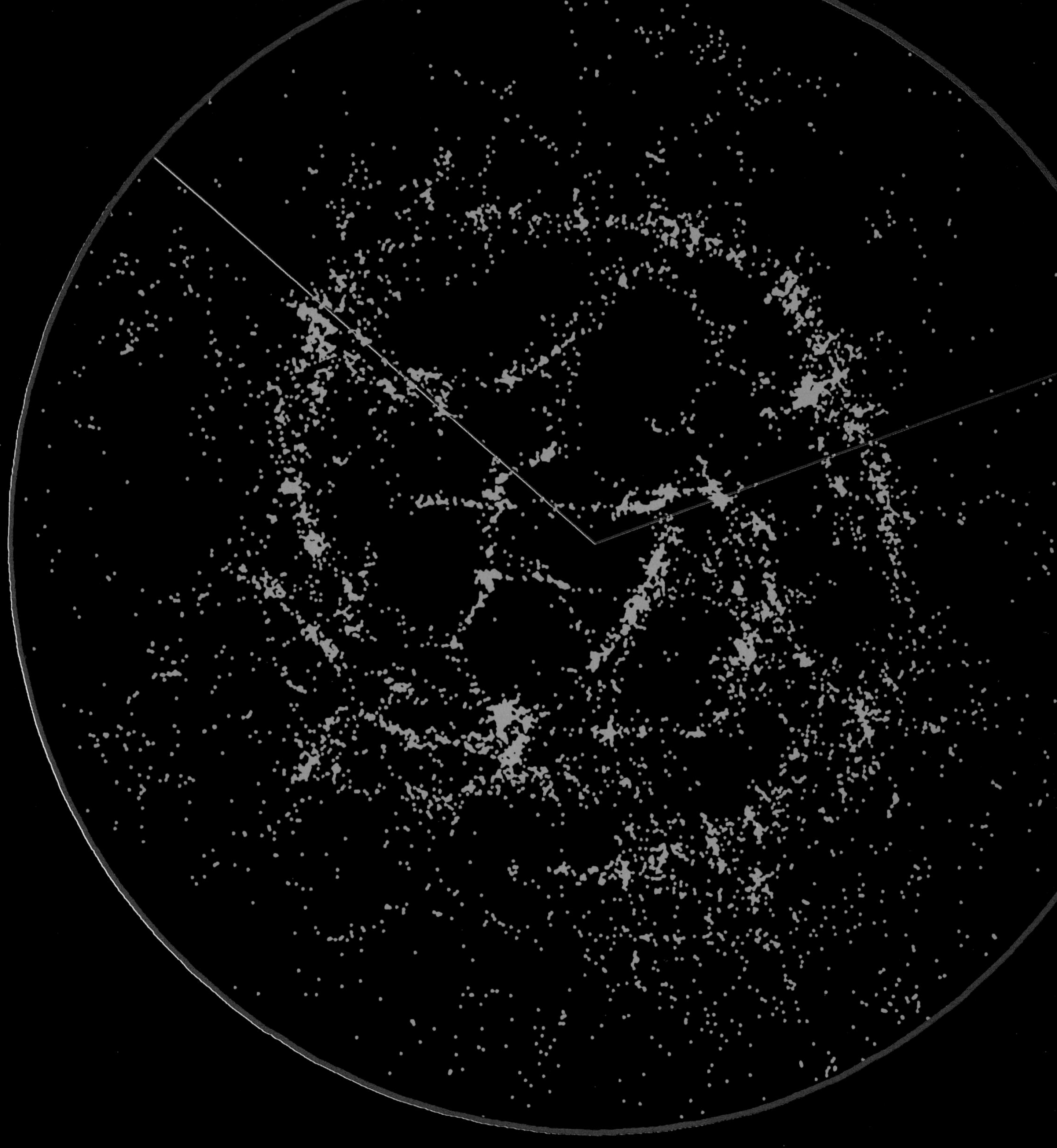

ergy in the early universe, you would expect to find distortions in the microwave background. Yet none have been found. That's perfectly consistent with gravity doing the job."

The Cold Dark Matter Solution

One of Gott's graduate students, Changbom Park, recently developed a supercomputer simulation that demonstrates how gravity alone can forge large-scale structure in an initially smooth universe. Park's simulation is based on the extreme form of dark matter called cold dark matter. Cold dark matter is simply a generic name given to a hypothetical group of massive, slow-moving (hence cold) particles such as gravitons, axions, and photinos. Because the cold dark matter particles are heavy and move slowly, much slower than the speed of light, they can be more easily bound together by gravity than "hotter" particles like neutrinos, which are theorized to make up cold dark matter's counterpart, hot dark matter.

The cold dark matter model predicts that small density irregularities form just 10^{-35} second (ten billionths of a billionth of a billionth of a billionth of a second) after the Big Bang. The irregularities arise from the random motions of the particles as described by the uncertainty principle, which states that we can never know how many particles exist in a certain region of space at any one time. On the quantum level, then, the particles can suddenly amass more in one place than another, forming wrinkles in the otherwise smooth distribution of matter. Because they develop in the sluggish

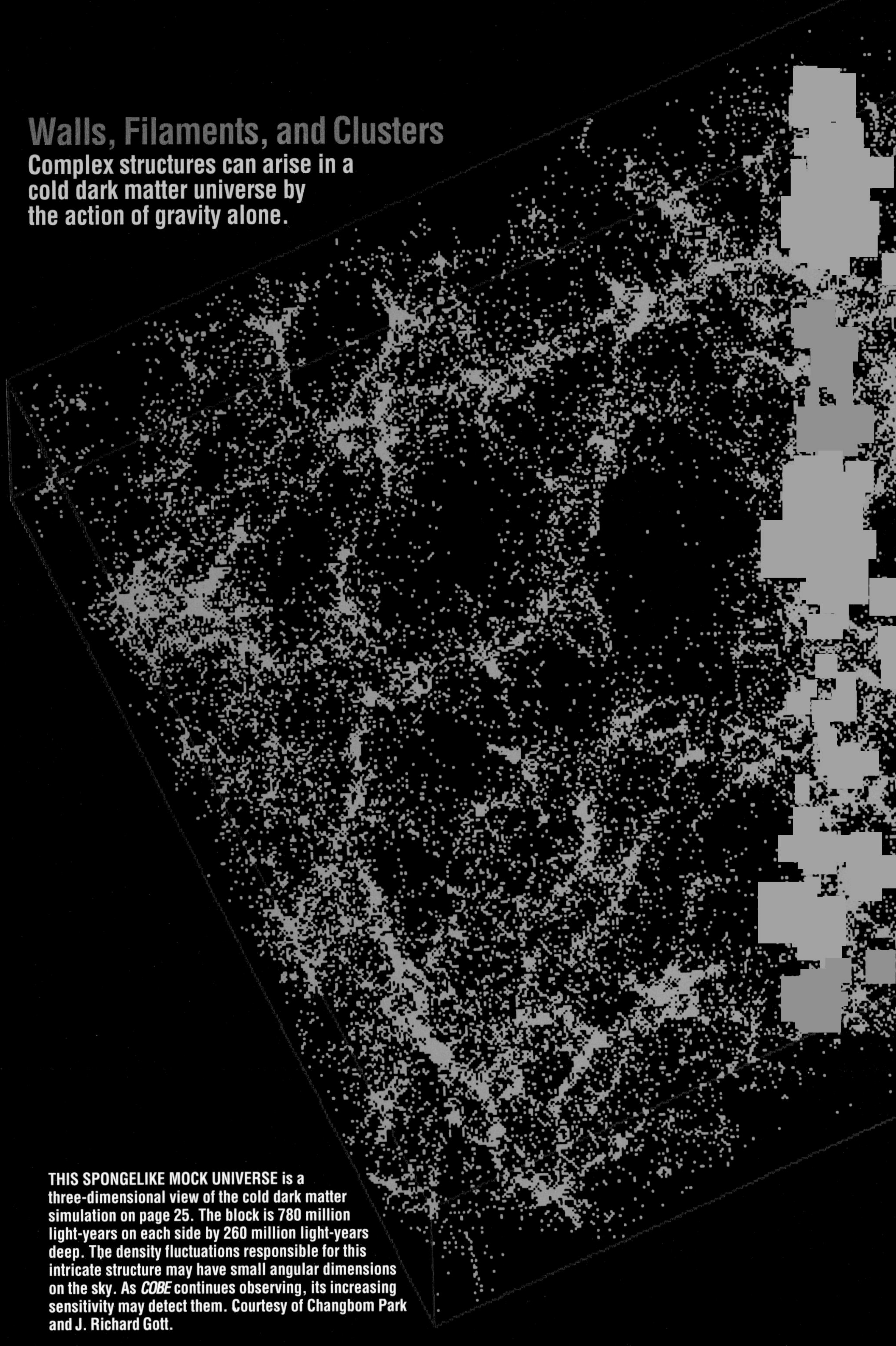

Walls, Filaments, and Clusters

Complex structures can arise in a cold dark matter universe by the action of gravity alone.

THIS SPONGELIKE MOCK UNIVERSE is a three-dimensional view of the cold dark matter simulation on page 25. The block is 780 million light-years on each side by 260 million light-years deep. The density fluctuations responsible for this intricate structure may have small angular dimensions on the sky. As *COBE* continues observing, its increasing sensitivity may detect them. Courtesy of Changbom Park and J. Richard Gott.

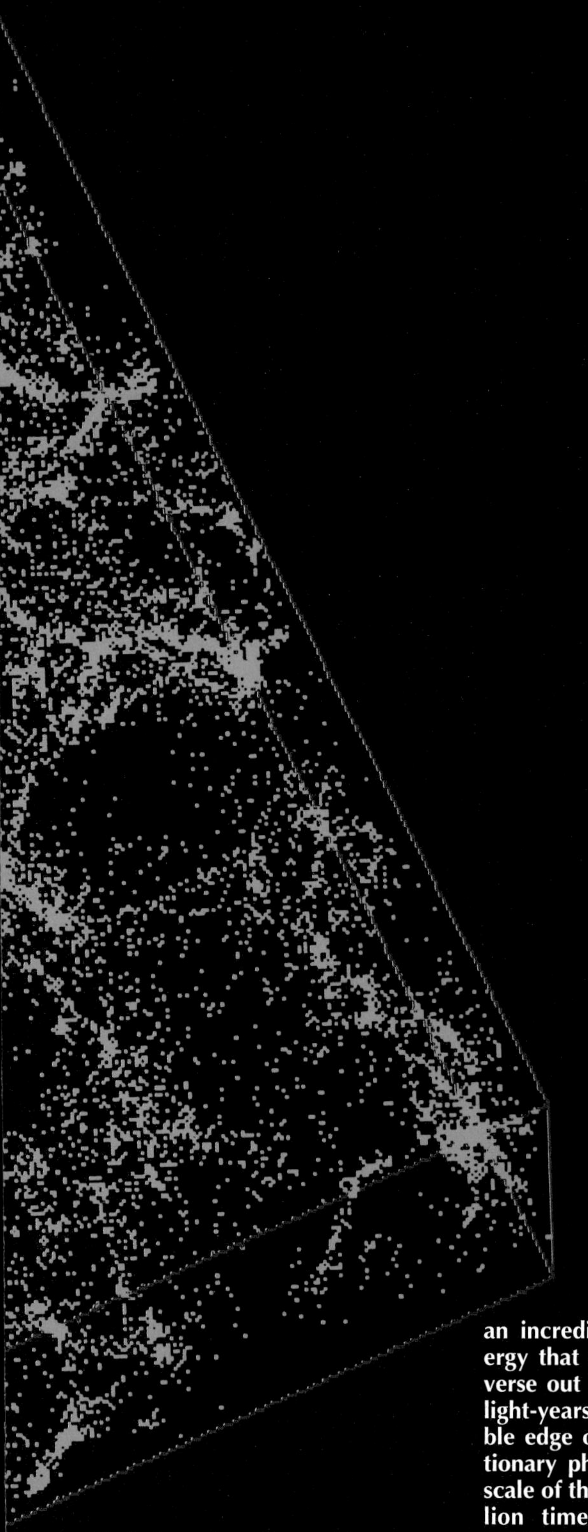

cold dark matter, the wrinkles can grow very soon after the Big Bang.

According to the inflationary model, 10^{-35} second after the Big Bang is a pivotal moment in the history of the universe, for at about that time the universe cooled enough to allow the primordial matter to "freeze" out into more normal matter, like quarks and electrons. This fundamental change in the state of matter — similar to water changing to ice — results in the release of an incredible amount of latent energy that suddenly inflates the universe out to hundreds of billions of light-years, well beyond the observable edge of our universe. The inflationary phase not only extends the scale of the universe ten trillion, trillion times, it also amplifies the original quantum irregularities to macroscopic scales. These congenital wrinkles, like a pattern on a deflated balloon, expand with the universe. By the time the universe is about 10,000 years old, they begin to grow into bigger wrinkles via gravitational instability.

The amplitude of these initial fluctuations are at first very small and can be directly calculated, said Gott. But when the universe is about 110 million years old, the fluctuations become too large to follow [illegible] simple mathematical formulae. [illegible]hat's where Park's simulation [illegible] into play. On a supercompute[illegible] the [illegible]gram plots the dis[illegible] [illegible]4 million particles, ha[illegible] [illegible] resent galaxies and [illegible]her half dark matter, and [illegible] how the distribution looks aft[illegible] [illegible]billion years if gravity were [illegible] force influencing structure. [illegible]verse that emerges looks striking[illegible] [illegible]milar to the one observed toda[illegible] [illegible]uding vast voids and g[illegible] [illegible]lls and clumps of galaxies.

Park's simulations, [illegible], are in accord with the *COBE* [illegible]ts so far. "I think it's important to remember that the cold dark matte[illegible]del predicted what *COB[illegible]* saw[illegible] said Gott. "The *COBE* grou[illegible] has found nothing down to 1 par[illegible] [illegible]0,000. That means the micro[illegible] background is very uni[illegible]he epoch when the unive[illegible] about 400,000 years old. [illegible] [illegible]ark matter model, howe[illegible] [illegible] fluctuations on the ord[illegible] 1 [illegible] in 200,000. That's smalle[illegible] [illegible] the limit of 1 in 10,000, so *COB[illegible]* [illegible]d not have been expected to [illegible] the fluctuations yet."

The initial *COB[illegible]* results were based on measurem[illegible]ts of the sky at an angular scale o[illegible] [illegible]egrees. But as *COBE* reobserv[illegible] [illegible] the sensitivity of the d[illegible] [illegible]re-spondingly increase[illegible] Mather hinted that *COBE* m[illegible] [illegible]act have already reached th[illegible] [illegible]par[illegible] in 200,000 limit. "It will [illegible]bly take us about a year to b[illegible] confident of our results," he said[illegible] but w[illegible] may already have some parts of [illegible] sky that are that good."

With still a year and a half [illegible] observations remaining in its mission, *COBE* has already [illegible]tinguished itself by confirming [illegible] t[illegible]rse began with a hot Big Bang a[illegible]i-formly expanded an[illegible]d [illegible] its present state. Moreo[illegible] isothermal 2.7-kelvins backgroun[illegible] its unvarying nature severely c[illegible]rain the initial condition[illegible] necessa[illegible]r galaxy formation in the early universe.

For now, the cold dark matter model co-exists with the *COB[illegible]* observations, but the model's parameters could chang[illegible] [illegible]ficantly before the mission [illegible] some-where around the mag[illegible]itude of 1 in 200,000, the un[illegible]ontinues to look smooth[illegible]ologists will have to resort to [illegible]g galaxies out of their theoreti[illegible]ts. *COBE* may have found the perfect universe, but it remains to be seen how simple or complex that perfection is. □

Observations of the early universe made by the Cosmic Background Explorer satellite reveal the seeds that grew into the galaxies and galaxy clusters we see today.

by Richard Talcott

Structure pervades the universe. Everywhere astronomers look, they see it. Anyone who goes outside on a clear night and gazes at the sky can see the most obvious structure. The planets and stars — distinct points of light surrounded by the blackness of space — confirm that matter clumps together more in some regions than in others.

As astronomers probed deeper into space this century, they found new levels of structure. Stars clumped into galaxies, and galaxies clumped into clusters. Within just the last two decades, astronomers have discovered that the clusters of galaxies themselves clustered into superclusters, which are separated by huge and apparently empty voids.

Structure seemed to be everywhere, except at the largest and most distant scale of all. When astronomers observed the cosmic microwave background — the remnant radiation from the big bang that gave birth to the universe — it appeared extraordinarily smooth. There was no hint of any structure at all. Everyone agreed that there had to be some irregularities in the otherwise featureless background, some seeds around which gravity could pull in matter to form the galaxies, but no one could find them. Until now.

In late April, a team of astronomers led by George Smoot of the Lawrence Berkeley Laboratory and the University of California announced that the Cosmic Background Explorer satellite (COBE for short) had discovered tiny variations in the temperature of the microwave background. According to Smoot, "These small variations are the imprints of tiny ripples in the fabric of space-time put there by the primeval explosion. Over billions of years, the smaller of these ripples have grown into galaxies, clusters of galaxies, and the great voids in space."

An Historic Finding

Speaking for many astronomers, University of Pennsylvania cosmologist Paul Steinhardt stated that if this finding is confirmed, it would rank as "not only one of the most important discoveries in cosmology, but one of the most important discoveries in science in this century." Yet this momentous discovery is based on just the tiniest variations.

The ripples are actually minuscule differences in the temperature of the background radiation. Embedded in the background radiation, which has a temperature of 2.73 kelvins (2.73° Celsius above absolute zero), the ripples represent deviations of just 30 millionths of a degree. That's equal to 1 part in 100,000 of the total signal.

The ripples appeared in observations made with COBE's Differential Microwave Radiometers. The six radiometers have been observing the sky continually since COBE's launch in November 1989. The discovery comes from data collected during the first year of the satellite's life. In that time, each of the six radiometers collected roughly 70 million measurements, for a total of nearly half a billion measurements. Then the computers went to work. Alan Kogut, a team member at Goddard Space Flight Center, says "Each of those measurements is like one piece of a gigantic jigsaw puzzle — you look at the piece by itself and it could mean anything. It's only when you fit all 70 million pieces for each of the radiometers together that the pattern starts to emerge."

The astronomers analyzed these measurements statistically to get the results. Although the statistical nature of the analysis means you can't point to any specific area of the sky and say, "Yes, this spot is one part in a hundred thousand hotter than that spot over there," it is a tried-and-true method for getting more sensitivity out of an instrument.

Team members didn't announce results immediately, however. They kept it to themselves for a year or more so the data and the computer programs

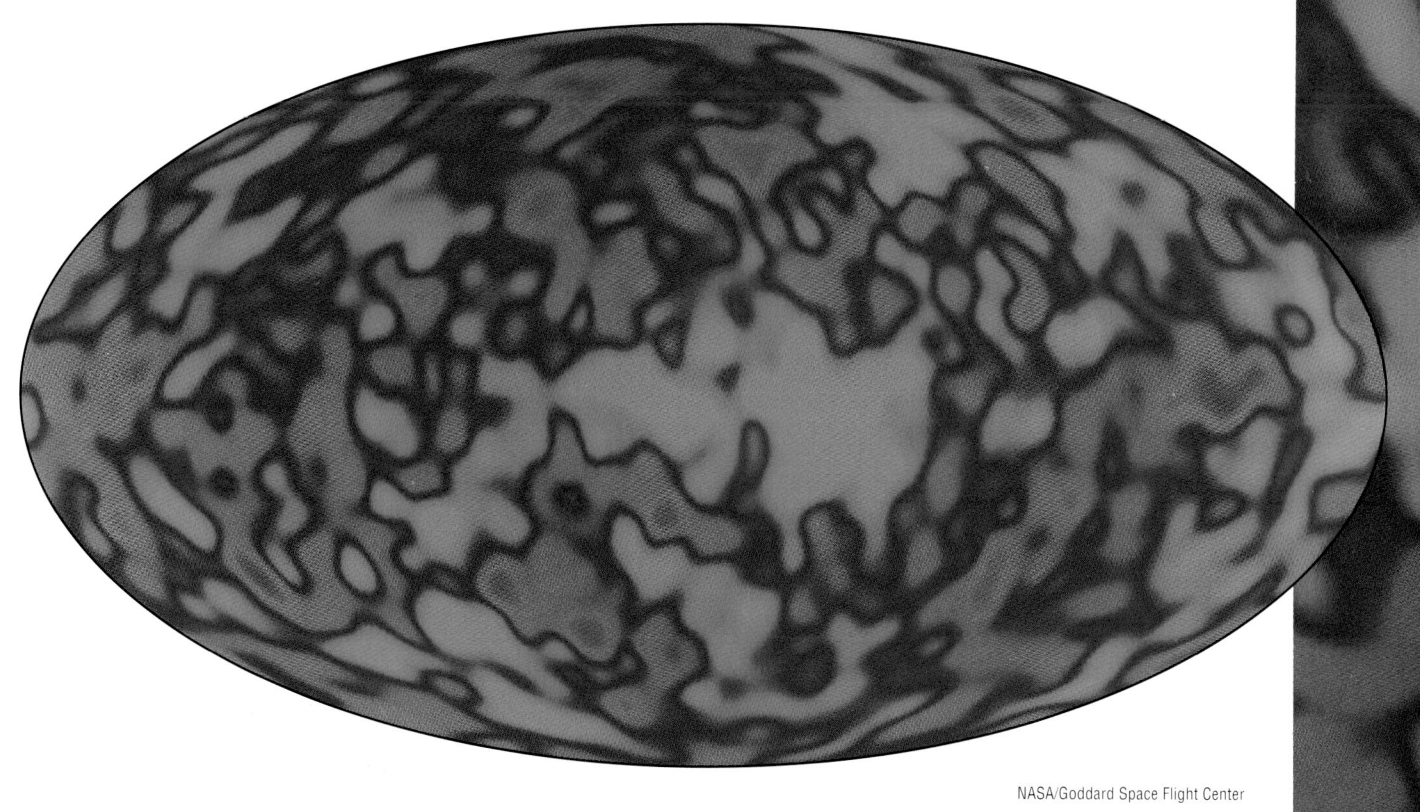

NASA/Goddard Space Flight Center

SEEDS OF STRUCTURE permeate the microwave background (red indicates variations 0.01% warmer and blue 0.01% cooler than the average 2.73 kelvins). Instrument noise causes most of the patchiness shown, but statistical analysis also indicates the presence of tiny — 0.001% — cosmic signals.

could be checked, re-checked, and checked again. They wanted to be sure they were seeing a cosmic signal and not a spurious one. And now, they are as sure as they can be.

What It Means

The discovery is far more than a single piece of the cosmological puzzle. It provides direct support for the theory of inflation, which proposes that the universe expanded extraordinarily fast for a brief period just after the big bang. It also supports the cold dark matter model of how the universe's observed structure arose. And even more importantly, it eliminates a large number of competing models.

The big bang theory grew out of Edwin Hubble's discovery in the 1920s that the galaxies are flying apart from one another and that the farther away a galaxy is, the faster it moves. But if the universe were expanding as this suggested, then at some point in the distant past, all the galaxies would have been much closer together. The big bang theory states that the entire universe was once packed into a point of infinite density and extraordinarily high temperature. The big bang created not only the universe and all the matter in it, but also the space that makes up the universe and time itself.

As the universe expanded, it cooled. But for a long time, the temperature was too hot for atomic nuclei to capture electrons. Photons of light could travel only short distances before interacting with free electrons. Then, when the universe's temperature fell below 5,000 kelvins roughly 300,000 years after the big bang, atoms formed. With electrons tied up in atoms, radiation could travel freely, and the universe became transparent to light.

This is the light astronomers see when they look at the microwave background radiation. But the radiation no longer has a temperature of 5,000 kelvins. Instead, the expansion of the universe has cooled the radiation to just 2.73 kelvins.

Beautiful, but Incomplete

For such a simple theory, the big bang theory works amazingly well. It explains the expansion of the universe. It predicts the correct abundances of hydrogen and helium, the two most common elements in the universe, which formed in the first few minutes after the big bang. But perhaps most impressively, it predicts the microwave background radiation. When Bell Labs researchers Arno Penzias and Robert Wilson detected this remnant radiation in 1964, astronomers quickly hailed the big bang as the best game in town.

Despite its power, the big bang was too simple to be a complete theory. It offered no reasons for many of the observed properties of the universe. For example, as astronomers examined the microwave background radiation, they found it to be incredibly isotropic — it appeared to have the same temperature no matter what direction they looked. But the big bang theory couldn't produce this. There was no way material in one part of the universe could communicate with material in a distant part of the universe so that they could reach equilibrium.

Inflationary models easily explain how matter and radiation reached equilibrium. First developed by MIT physicist Alan Guth in 1980, the inflation theory proposes that shortly after the big bang (10^{-35} second, or 10 trillion trillion trillionths of a second, to be exact), the universe underwent an exponential expansion that increased its size enormously (by a factor of 10^{50}). Before that time, the universe was small enough that widely separated parts could reach equilibrium, and that equilibrium persisted after the inflationary epoch ended.

What motivated inflation, according to Steinhardt, was an *a posteriori* prediction, or "postdiction." In other words, we had observations that

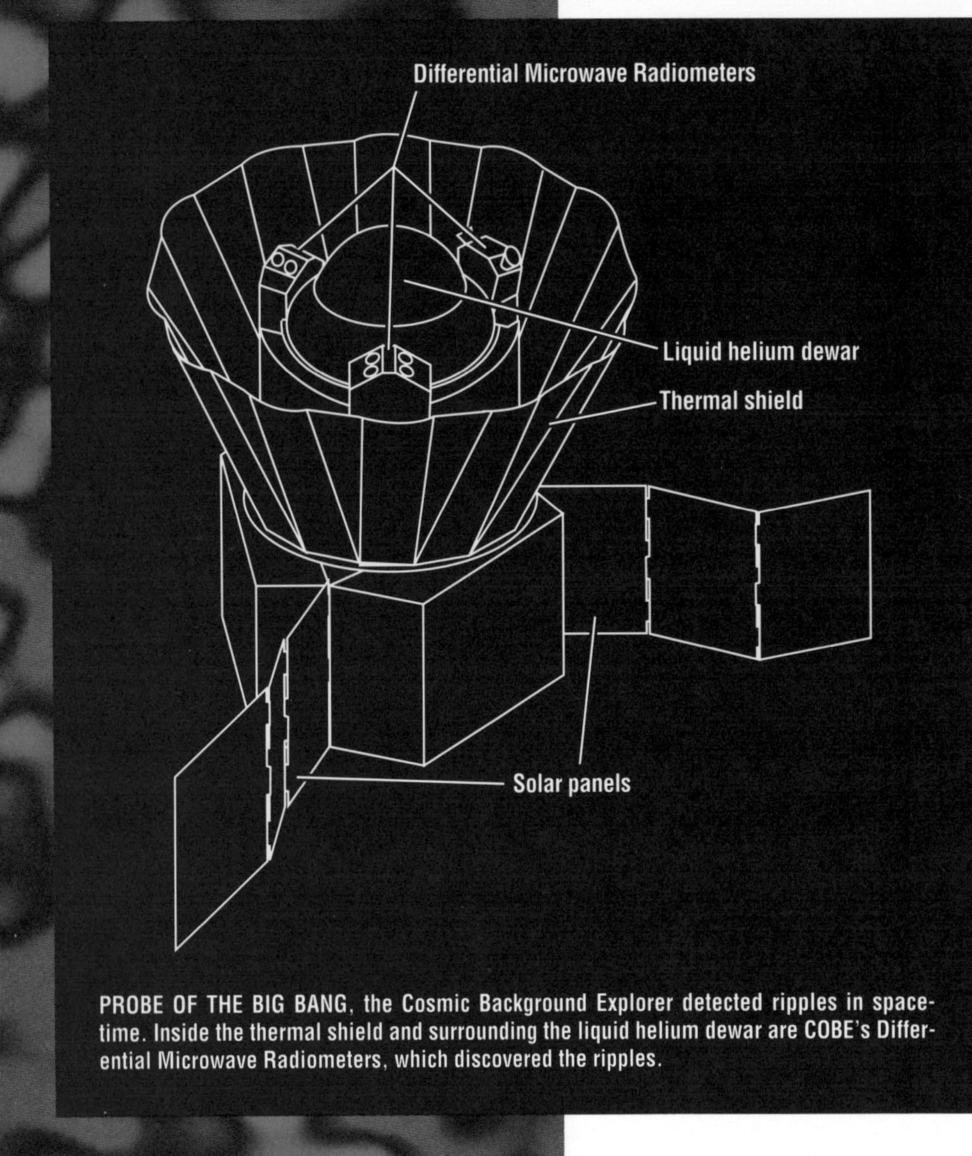

PROBE OF THE BIG BANG, the Cosmic Background Explorer detected ripples in space-time. Inside the thermal shield and surrounding the liquid helium dewar are COBE's Differential Microwave Radiometers, which discovered the ripples.

NASA

couldn't be explained under existing theory, so we developed a new theory that could explain them. Though inflation made predictions, no experimental confirmation existed.

As Steinhardt recalls, "A number of cosmologists got together to see whether any fluctuations might be left after inflation. And we discovered the surprising prediction that inflation leaves behind some inhomogeneities, or ripples, in the universe, and these ripples have a characteristic spectrum.

"The important thing about the COBE results is that the ripples agree with this spectrum, at least within the first limits that the researchers found. That's the first prediction, rather than a postdiction, made by inflation to be confirmed."

Wanted: Matter, Cold and Dark

COBE's results also support the contention that lots of "dark matter" exists in the universe. In fact, as much as 90 percent or more of the universe may be in the form of dark matter, which is unlike ordinary matter and can be seen only through its gravitational effect on ordinary matter. (See "Shedding Light on Dark Matter," February 1992.)

Smoot says the temperature and size of the variations discovered by COBE agree with theories of dark matter. "Theory tells us that these vast regions cannot be composed of ordinary matter. It must be something we've never seen in our laboratories. If it were ordinary matter, which interacts with light, then we would see much greater temperature variation. Only matter that does not interact with light except through gravity could cause such slight temperature variation."

Dark matter can take many forms, but the COBE results seem to point toward cold dark matter as the best candidate. Cold dark matter consists of slow-moving, massive particles that can gravitationally bind together fairly easily to form the seeds that subsequently draw in ordinary matter. Other possible seeds bandied about by theorists include hot dark matter (faster moving cousins of cold dark matter) and various defects in the topology of space called textures, cosmic strings, and domain walls.

Paul Steinhardt explains the significance of COBE's findings: "What COBE has done has, first of all, produced results that are consistent with the basic outlines of inflation and are consistent with the cold dark matter picture. And, as it now stands, COBE has wiped out the texture scenario entirely. It's also wiped out the domain wall scenario. And finally, it's wiped out all of the cosmic string scenarios except for those where you have strings plus hot dark matter."

What the Future Holds

The COBE results are impressive, but everyone involved cautions that they need to be confirmed. That confirmation may come sooner rather than later. Balloon-borne experiments and telescopes at the south pole are looking for these same variations and achieve nearly the same level of sensitivity as COBE. In addition, there's another year's worth of COBE data for Smoot's team to analyze.

The confirmation may come within the next year or so. Steinhardt thinks that within a year we'll either see that something has fooled COBE or have confirmation. If it is confirmed, Smoot says, "Our results will unify physics on the largest and smallest scales, fusing together the fields of particle physics and cosmology." When that day comes, we'll all know a lot more about how our universe got to be the way it is. □

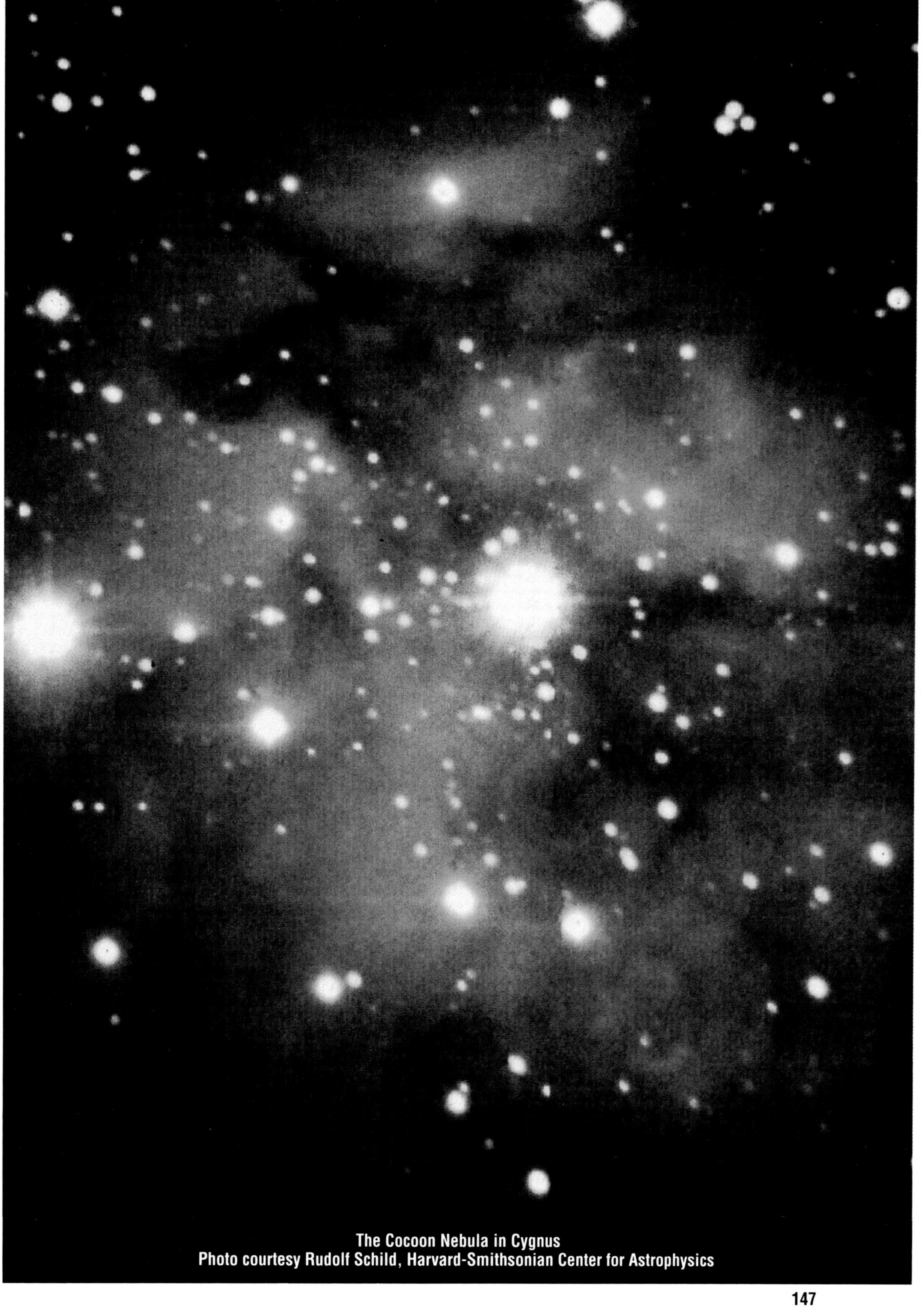

The Cocoon Nebula in Cygnus
Photo courtesy Rudolf Schild, Harvard-Smithsonian Center for Astrophysics

Peculiar Galaxy IC 356 in Camelopardalis
Photo courtesy Rudolf Schild, Harvard-Smithsonian Center for Astrophysics

Bibliography

Allen, David A. *Infrared; the new astronomy.* 228 pp., hardcover. John Wiley & Sons, New York, 1975.

Bok, Bart J., and Priscilla F. Bok. *The Milky Way.* Fifth ed., 356 pp., hardcover. Harvard University Press, Cambridge, Massachusetts, 1981.

Cadogan, Peter. *From Quark to Quasar; notes on the scale of the universe.* 183 pp., hardcover. Cambridge University Press, New York, 1985.

Culhane, J. Leonard, and Peter W. Sanford. *X-Ray Astronomy.* 192 pp., hardcover. Charles Scribner's Sons, New York, 1981.

Davis, Joel. *Journey to the Center of Our Galaxy; a voyage in space and time.* 335 pp., hardcover. Contemporary Books, Chicago, 1991.

Ferris, Timothy. *Coming of Age in the Milky Way.* 495 pp., hardcover. William Morrow, New York, 1988.

Ferris, Timothy. *Galaxies.* 191 pp., hardcover. Sierra Club Books, San Francisco, 1980.

Ferris, Timothy. *The Red Limit; the search for the edge of the universe.* 286 pp., paper. Quill, New York, 1983.

Field, George, and Donald Goldsmith. *The Space Telescope; eyes above the atmosphere.* 276 pp., hardcover. Contemporary Books, Chicago, 1989.

Friedman, Herbert. *The Astronomer's Universe; stars, galaxies, and cosmos.* 359 pp., hardcover. W.W. Norton, New York, 1990.

Goldsmith, Donald. *The Astronomers.* 332 pp., hardcover. St. Martin's Press, New York, 1991.

Harwit, Martin. *Cosmic Discovery; the search, scope, and heritage of astronomy.* 334 pp., hardcover. Basic Books, New York, 1981.

Hawking, Stephen W. *A Brief History of Time, from the Big Bang to black holes.* 198 pp., hardcover. Bantam Books, New York, 1988.

Hodge, Paul. *Galaxies.* 174 pp., hardcover. Harvard University Press, Cambridge, Massachusetts, 1986.

Jaschek, Carlos, and Mercedes Jaschek. *The Classification of Stars.* 413 pp., hardcover. Cambridge University Press, New York, 1987.

Kaufmann, William J. III. *Galaxies and Quasars.* 226 pp., hardcover. W.H. Freeman and Co., New York, 1979.

Laustsen, Svend, Claus Madsen, and Richard M. West. *Exploring the Southern Sky; a pictorial atlas from the European Southern Observatory (ESO).* 274 pp., hardcover. Springer-Verlag, New York, 1987.

Malin, David, and Paul Murdin. *Colours of the Stars.* 198 pp., hardcover. Cambridge University Press, New York, 1984.

McDonough, Thomas R. *The Search for Extraterrestrial Intelligence; listening for* *life in the cosmos.* 244 pp., hardcover. John Wiley & Sons, New York, 1987.

Mitton, Simon. *The Crab Nebula.* 194 pp., hardcover. Charles Scribner's Sons, New York, 1978.

Murdin, Paul, and Leslie Murdin. *Supernovae.* 185 pp., hardcover. Cambridge University Press, New York, 1985.

Overbye, Dennis. *Lonely Hearts of the Cosmos; the story of the scientific quest for the secret of the universe.* 438 pp., hardcover. HarperCollins, New York, 1991.

Riordan, Michael, and David N. Schramm. *The Shadows of Creation; dark matter and the structure of the universe.* 277 pp., hardcover. W.H. Freeman and Co., New York, 1991.

Sagan, Carl. *Cosmos.* 365 pp., hardcover. Random House, New York, 1980.

Silk, Joseph. *The Big Bang.* 485 pp., hardcover. W.H. Freeman and Co., New York, 1980.

Tucker, Wallace, and Riccardo Giacconi. *The X-Ray Universe.* 201 pp., hardcover. Harvard University Press, Cambridge, Massachusetts, 1985.

Verschuur, Gerrit L. *Interstellar Matters.* 320 pp., hardcover. Springer-Verlag, New York, 1988.

Verschuur, Gerrit L. *The Invisible Universe Revealed; the story of radio astronomy.* 262 pp., hardcover. Springer-Verlag, New York, 1987.

Weinberg, Steven. *The First Three Minutes; a modern view of the origin of the universe.* 188 pp., hardcover. Basic Books, New York, 1977.

Spiral Galaxy NGC 3521 in Leo
Photo courtesy Rudolf Schild, Harvard-Smithsonian Center for Astrophysics

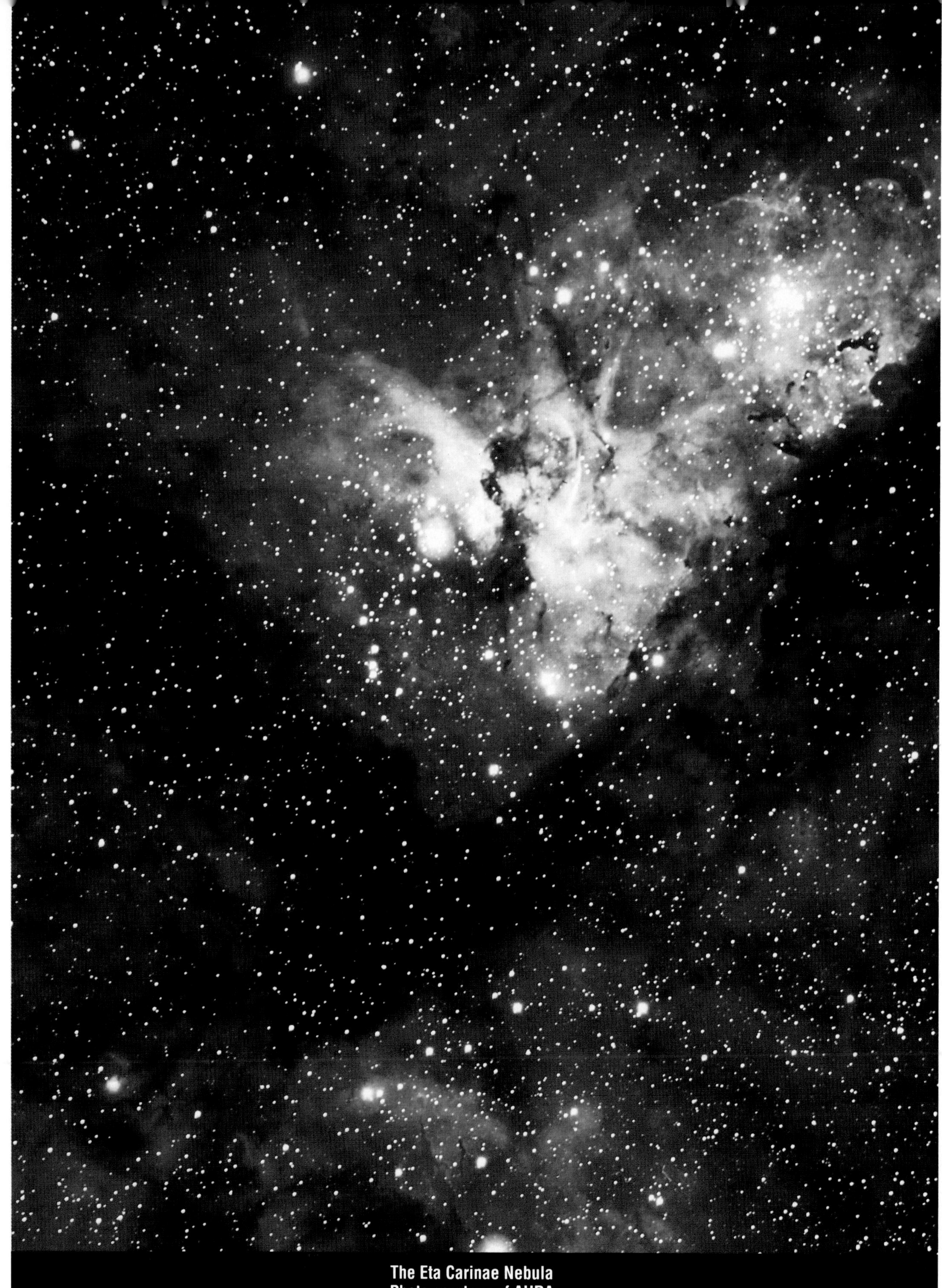

The Eta Carinae Nebula
Photo courtesy of AURA

The Authors

Alan P. Boss is an astrophysicist at the Department of Terrestrial Magnetism of the Carnegie Institution of Washington, Washington, D.C.

Kenneth Brecher is a professor of astronomy and physics at Boston University in Boston, Massachusetts and director of the Boston University Science and Mathematics Education Center.

Robert Burnham is editor of ASTRONOMY Magazine, having served on the editorial staff since 1978. He is the author of *The Star Book* (AstroMedia and Cambridge University Press, 1985) and adds an active role as an observer to his experience as a science journalist.

Jack O. Burns chairs the astronomy department at New Mexico State University. He is also an adjunct researcher at the National Radio Astronomy Observatory and has been a long-term visiting scientist at the University of Illinois' National Center for Supercomputing Applications.

Ken Croswell received his doctorate in astronomy from Harvard University and has written for many publications, including ASTRONOMY, *Astronomy Now, New Scientist, Science, Star Date,* and Time-Life Books.

David J. Eicher is associate editor of ASTRONOMY Magazine, where he has worked since 1982. He is the founder and former editor of *Deep Sky,* and has written, edited, or contributed to six astronomy books, including *Stars and Galaxies; ASTRONOMY's guide to exploring the cosmos* (AstroMedia, 1992).

Peter Jedicke is an enthusiastic amateur astronomer who teaches math and physics at Fanshaw College in London, Ontario.

Jeff Kanipe is a former associate editor of ASTRONOMY who now serves as editor of *Star Date* magazine, published by the University of Texas at Austin. He is an enthusiastic observer who enjoys both solar system and deep-sky objects.

David H. Levy is a freelance author and comet hunter, with seventeen comet discoveries to his credit. He is the author of *Observing Variable Stars; a guide for the beginner* (Cambridge University Press, 1989), and *The Sky; a user's guide* (Cambridge University Press, 1991).

Laurence A. Marschall is a professor of physics at Gettysburg College in Gettysburg, Pennsylvania. He is the author of *The Supernova Story* (Plenum Press, 1988).

Richard Monda is director of the Schenectady Museum Planetarium in New York.

Govert Schilling is a regular contributor to the Dutch publication *Zenit* and an author of several astronomy books published in the Netherlands.

Richard Talcott has been an Assistant Editor at ASTRONOMY Magazine since 1986. His main interests are in stellar and extragalactic astronomy, cosmology, and computer mapping of the heavens.

Gerrit L. Verschuur is a radio astronomer and freelance writer living near Bowie, Maryland. He is the author of several books, including *The Invisible Universe Revealed; the story of radio astronomy* (Springer-Verlag, 1987), and *Interstellar Matters* (Springer-Verlag, 1988).

Belinda J. Wilkes is an astrophysicist at the Harvard-Smithsonian Center for Astrophysics in Cambridge, Massachusetts.

Distant Barred Spiral Galaxy NGC 4762 in Virgo
Photo courtesy Rudolf Schild, Harvard-Smithsonian Center for Astrophysics

Index

Note: Photographs, diagrams, and illustrations are in bold type.